MAN MANAGEMENT
and Rig Management

ROTARY DRILLING • UNIT IV

Edited by Jean Pietrobono

Published by

PETROLEUM EXTENSION SERVICE
The University of Texas at Austin
Continuing and Innovative Education
Austin, Texas

Originally produced by

INTERNATIONAL ASSOCIATION
OF DRILLING CONTRACTORS
Houston, Texas
1987

Published 1987. Sixth Impression 2012
Printed in the United States of America

Catalog no. 2.40000
ISBN 0-88698-128-X

No state tax funds were used to publish this book. The University of Texas at Austin is an equal opportunity employer.

Contents

Preface

In 1981, Petroleum Extension Service was charged by the International Association of Drilling Contractors with producing a nine-part video series about personnel management for first-line supervisors in the oil patch. That series roughly tracked the original Unit IV of the Rotary Drilling Series. Because many drilling contractors and other companies in the industry found the video series valuable, IADC urged PETEX to revise the text upon which it was based. The original unit was published in eight separate booklets. PETEX found it more efficient to put all eight lessons under one cover for the present volume.

This edition of *Man Management and Rig Management* draws heavily from research done by M. L. Powers for the original lessons. PETEX is greatly indebted to the Man Management and Rig Management Task Group of the IADC Training Subcommittee, under the Human Resources Committee, for its work on this edition. Members of the Task Group are Jan Pedersen of ODECO, Task Group chairman; Roger Cumba of Diamond M Company, Training Subcommittee chairman; Bill Person of Rowan Companies, Human Resources Committee chairman; Lee Hunt of IADC; Ron Weaver of Atwood Oceanics; Dennis Luce of Pool Offshore Company; Hoss Ramke of Teledyne Movible Offshore; Ron Ide of IADC; Lonnie Leger of Blocker Energy; Jack Gaffron of Conoco; Caroll and Mary Jon Eubanks; and Ron Baker of Petroleum Extension Service. Special thanks also are extended to Louis Mullenix of Penrod Drilling Company for his helpful critique of the text.

How to use this manual

A set of objectives is given at the start of each of the eight lessons in this manual. Read the objectives carefully before studying the material in each lesson. As you study, take notes using the objectives as a guide to the most important parts. The self-test in the center of the manual covers all eight lessons. The test is based on the lesson objectives. Since it is a self-test, you must decide whether you should refer back to the material to answer the questions. An answer key is provided at the back of the manual.

Lesson 1

WHAT IS MANAGEMENT?

OBJECTIVES

Upon completing this lesson, the student will be able to

- ☑ *Identify the five interests in any business and describe the concerns of each.*
- ☑ *Identify which people in a company are management.*
- ☑ *Discuss the three basic management skills: planning, instructing, and coordinating.*
- ☑ *Explain the two steps involved in planning.*
- ☑ *Discuss the kinds of instructing that a manager does.*
- ☑ *Describe how a manager coordinates work.*

The question posed by this lesson can be answered simply: Management is the people who manage or direct a business. To get at what this means, this lesson takes a look first at why we need management, then who management is, and finally what management does.

The reason we need management lies in the nature of business—that's any business, drilling included. A business exists (1) to provide a product or service and (2) to make a profit. Five interests are involved in the pursuit of those goals: the investor, the worker, the consumer, the competitor, and the government. The purpose of management is to balance the needs of those five interests in the profitable operation of a business.

Management is anyone in a business who is responsible for the work of others. In a typical drilling business, that includes everyone from the driller on up to the company president. Everyone in management is concerned with meeting the company's ultimate goal of running a sound business.

Managers do three basic things: plan, instruct, and coordinate. Mastery of these three skills is essential to good management. These skills will be discussed often in the lessons that follow.

WHY DO WE NEED MANAGEMENT?

The five interests of any business

To understand why a business needs management, we must first consider what a business is. A business is where you work a certain number of hours each week to earn money. Right? But a business involves more people than you and the person who signs your paycheck. For example, somebody has to put up the money for the tools and materials you need to do your job and has to provide a place to work. He is the investor in the business. And someone else must be willing to pay for whatever it is that you produce or for the service you provide. That someone else is the customer, or consumer.

So in addition to you, the worker, a business also involves the investor who puts up the front money and the consumer who buys what you produce. In almost any business, two other interests are involved as well: the competitor and the government.

Consider the roles of these five interests in a simple business—say a popcorn vendor. The investor puts up money to buy a corn-popping machine, oil, salt, and the corn. Workers are hired to do the popping. Consumers, of course, buy and eat the popcorn. The popcorn must be fresh and priced fairly, so that the consumers don't choose to buy instead from the popcorn stand down the road, the competitor. The government levies taxes on the popcorn vendor and also makes sure that working conditions are acceptable and the popcorn is safe to eat.

If a business is going to succeed, the interests of all five of these groups have to be balanced. That's the purpose of management, namely, seeing to it that the needs of all five of these groups in a business are met—the investor, the worker, the consumer, the competitor, and the government. Let's look more closely at the needs of each of these five interests, that is, what they expect to get out of the business, and what management must do to see that they get it.

The investor

When an investor puts money into a business, he expects two basic things: (1) a return on his investment, that is, a profit, and (2) security, or reasonable assurance that he won't lose his investment. The relationship between security and return is inverse: the greater the security, the lower the return.

Say, for example, that you have $10,000 to invest. If security is your highest priority, then you probably would put the money in a savings account or in savings bonds. Both of these options are safe. There is little chance that you will lose your $10,000 with either of them. Your rate of return, however, would be fairly low. On the other hand, if you're willing to sacrifice some security in hopes of higher returns, then you might consider investing your $10,000 in stocks or in the drilling business, where risks are higher but so are potential profits.

The worker

Why do people work? For money and security, of course. But workers have many other important needs beyond these, such as a sense of belonging to the organization and the opportunity to advance. Keep in mind that people's needs vary. For some people, security is most important, and they might be willing to accept a lower-paying or less-interesting job if they knew that they would never be fired from it. Other people would rather take a less-stable job for more money, wider recognition by the public, or power in the company.

Five interests are involved in any business: the investor, the worker, the consumer, the competitor, and the government.

The consumer

The two primary concerns of the consumer are price and quality. Sometimes price will take precedence over quality; other times it is the other way around. If quality is equal, however, the consumer will seek the lowest price. This is true of consumers buying a product or service from any business, including the drilling business. When a consumer submits a well for bids, it usually goes to the lowest responsible bidder. Note the word *responsible.* If a driller is not responsible, that is, if he doesn't do quality work, then the consumer may choose a higher bidder anyway. In this case, the consumer would choose quality over price.

The competitor

The competitor's goal is to lure away as many of your customers as he can. He will attempt to do this by offering a product that is the same or better than yours at the same or lower price. Competitors exist in any business; keeping ahead of them is crucial.

The government

The government's most visible concern with a business is its share of your profits, which it collects primarily through income taxes. Government also affects business through regulation of areas such as wages and hours, safety, and production allowables.

What must management do to meet the needs of each of these five interests? First, management must run the business efficiently so that it stays afloat and makes a profit for the investor. A big part of this is making sure that workers perform their jobs the best that they can. Management also must see to it that the business offers a quality product or service at a fair price; if not, customers will turn to the competitor. Finally, management must also make sure that all governmental regulations are followed.

WHO IS MANAGEMENT?

Not the mysterious "they"

To most people in a company or organization, management is a mysterious thing or someone, often referred to only as "they." But who, really, are "they"? Consider the organizational chart of a typical small drilling company. At the top is the president, with the vice-president below. Next is the superintendent in charge of field operations, followed by the toolpusher, driller, and crew. The administrative side of the organization has a similar managerial chain. Which of these people are management? Some would say only the president and vice-president are managers. Others might draw the line at field superintendent.

GODEEP DRILLING COMPANY ORGANIZATIONAL CHART

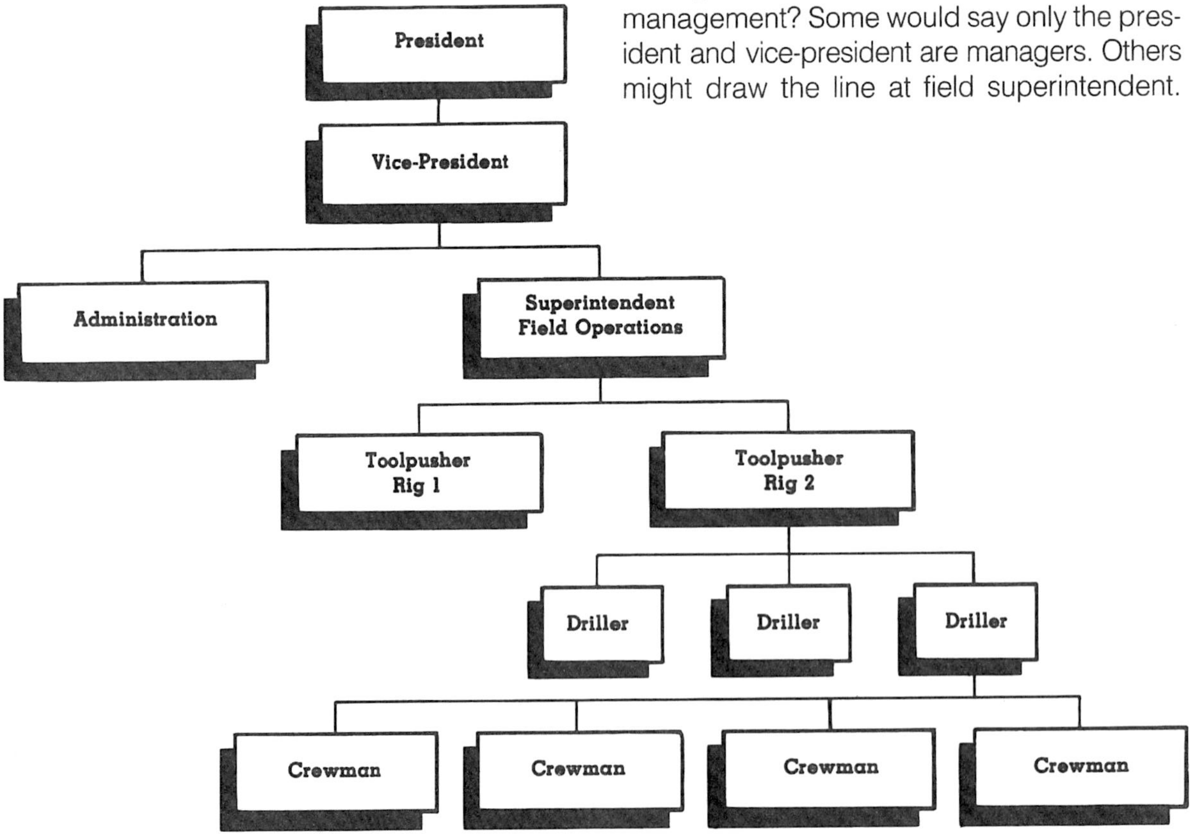

In a typical small drilling company, management includes everybody from the driller on up.

But management includes anyone in an organization who is responsible for the work of others. On this chart, that would include everyone from the driller, who is responsible for the work of his crew, on up.

Most larger companies have many investors: the stockholders who buy shares in the business. A board of directors looks after the interests of the stockholders. In such companies, management can be thought of as a chain tying together the board of directors on one end and the workers on the other. The board of directors must see to it that the funds invested by the stockholders are used wisely so that the company grows and the stockholders receive dividends. For that to happen, the workers on the other end of the chain must perform their jobs. Management, in between, must make sure that everybody does their part to meet the ultimate goal of a profitable business. All links in the chain must pull together to reach that goal.

WHAT DOES MANAGEMENT DO?

The three basic skills

A group of managers was asked to write down what they do on the job. Here is a partial list of what they came up with: plan work, give orders, see that work gets done, hire new hands, keep up discipline, motivate workers, keep records, make reports, conform to government regulations, prevent accidents, train hands, take care of personnel problems, improve work methods, keep costs down, work with service and supply companies, and promote good public relations.

These are some of the things that a manager does. You could probably add many more jobs to the list. But all of these jobs, and the many others that a manager performs, can be grouped under three basic skills: planning, instructing, and coordinating. Mastery of these three skills is the secret to being a good manager.

Planning

Any kind of planning involves two steps: setting the objective and providing the means. Personnel development and keeping up to date on new technology are forms of long-term planning that are of particular importance to the manager.

Setting the objective. The first part of planning is deciding what must be done. Say, for example, you are the driller on a rig that has a mud pump that keeps leaking mud around one of the rods, and you are forced to shut down while you wait for the yard to bring out new rod packing. You decide that the best way to cut down on future delays like this would be to keep plenty of spare packing on the rig. That's setting an objective—the first step in planning.

Some managers do not like to set objectives because it involves making decisions and taking some risks. If they make the wrong decision, then they are responsible for the consequences. These managers would rather take orders from the supervisors above them. But managers at all levels should be involved in planning and setting objectives. The driller, who's on the rig every day, is the most likely person to be able to spot a problem in daily work and come up with a plan to solve it.

Providing the means. The second step in planning is to provide the means for reaching your objective. For the driller with the mud pump problem, that might be asking the yard to send out new packing before he needs the next replacement, and then seeing to it that more is sent out as needed.

Personnel development. Preparing workers for advancement is an important part of long-term planning. The benefits to a manager include better performance out of his people and a lower turnover rate. People will work harder if they know that doing so will help them to move up in the organization. People also will stay with an organization longer if they know that they have a chance for advancement.

Personnel development can be on the job. A driller, for example, could show his derrickman how to run the brake, so that when the driller is ready to move up to toolpusher, the derrickman can take over as driller. Personnel development also includes encouraging crew members to take outside training courses that would help them advance in their jobs.

Keeping up to date technically. Staying current on technological developments in the industry is another kind of planning that managers must do. If, for example, you're going to be installing a new BOP stack on your rig, you should take the initiative to find out as much as possible about it. That might mean reading the manufacturer's manual or talking with the supply company's representative. Technical short courses, correspondence courses, and night classes are a good way to stay informed on a wide range of subjects. Another way to keep abreast of changing technology is to read trade magazines. You also should make new technical information available to your workers, either on the job or by signing them up for schools or courses.

Reading up on new technology is an important part of planning.

Instructing

One kind of instructing is giving orders, that is, telling people what you want them to do. Giving orders is one of a manager's major responsibilities. But instructing also involves training—showing people how to do something and making sure they know how to do it the next time. Orienting new hands on the job and explaining company goals to workers are among the kinds of instructing that are a manager's responsibility.

Giving orders. Most managers must give orders to their workers every day. That's a big part of how you see to it that the work you are responsible for gets done. There are positive and negative ways to give orders, and the positive ways almost always bring about better results. That is, a worker who is given an order in a positive way will do the job better and more efficiently.

Tone of voice is very important in how orders will be received. If you give orders in a harsh or demeaning tone of voice, the worker's self-esteem and morale will be lowered. He might do whatever it is you told him to do, but nothing more. He also will likely lose some respect for you and also lose his sense of loyalty toward the company. If, however, orders are given in a pleasant, enthusiastic tone of voice, the worker will probably respond positively. He will be much more likely to do more than just the minimum, and also will probably feel better toward you and the company.

Another positive technique is to phrase an order as a request rather than as a demand. "Ron, could you tighten the bolts on the deadline anchor?" is much easier to take than "Tighten the bolts on the anchor now!" Requests are accepted more willingly by workers than demands. When workers accept orders willingly, they tend to have higher morale and to perform their work better.

Training. How good you are at training your people will largely determine how good you will be as a manager. If you train your crew right, they will perform their jobs right day in and day out, with minimal supervision. If they're not trained right, they will need constant help and leave you with little time for anything but retraining. A few of the basics of training:

- Before you begin to train a worker to do a particular job, find out what the worker already knows about it.
- Always explain why the job you are about to teach a worker is important. This helps motivate the worker.
- Explain each step involved in performing a job, from the most simple to the most complex.
- It is usually best to *show* a worker how to do a job as well as to *tell* him how.
- Allow a worker the chance to practice a new job many times until he is ready to perform it unsupervised.

Orienting a new hand. During the first few days on the job, a new hand forms opinions about the company, the crew, and his work. You should do your best to see to it that those opinions are positive, and that the new worker feels that he fits in with the organization. A manager's responsibilities in orienting a new hand usually include introducing him to other crew members, showing him around the rig, talking to him about what will be expected of him, and explaining company policies and safety rules.

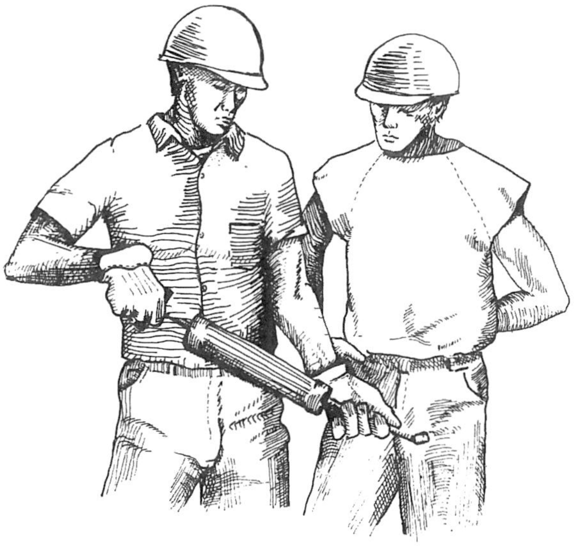

A good manager doesn't just* tell *how to do the job, he* shows *how to do it.

Explaining company goals to workers. Many companies don't tell workers about long-term goals, such as plans to buy a new rig or to move into a new territory. They often withhold information on the premise that workers don't need to know about it to do their jobs or that they wouldn't be interested in knowing about such things. But everybody likes to know what's going on; nobody likes to feel that they were the last to know. Workers are more likely to take pride in being part of their company, and to take pride in their work, if you keep them informed about what the company is doing or plans to do.

Coordinating

To a manager, coordinating means making sure that work is done as planned. It involves getting your crew to cooperate with each other and work together, following up on plans and making adjustments as necessary, and evaluating worker performance.

Getting the crew to cooperate and work together. Making hole is a team effort—everybody on the crew must work together to get the job done. The results of worker cooperation, or teamwork, are higher morale, higher productivity, fewer accidents and mistakes, and an easier job for the manager. You can promote teamwork by praising your workers when they do their jobs well, especially when they are working together. You should let the crew know that each of their jobs is important and that the rig couldn't make hole without them. Another way to encourage teamwork is to ask crew members for their suggestions and then consider them. They will take more pride in their work, and make a greater effort to work together, if they know that they have a say in how things are done.

Following up on plans. You cannot assume that work will automatically take place as planned. You must supervise your crew to see that the work is being done, and then be ready to change plans if necessary. For example, say that you've planned several tasks for a crewman, the last one being to change out the tong dies, before a trip out. Some unexpected delays arise as the day progresses, and it becomes clear that the crewman won't be able to change out the tong dies in time for the trip out. If you've been observing the crewman's work throughout the day, you would realize that he'd gotten behind and that he wouldn't be able to finish his work as planned. You must then be ready to adjust plans: Either get another crewman to change out the tong dies or delay the trip long enough to get them changed. You must be prepared to make these kinds of decisions daily.

Praising a worker for a job well done helps to promote teamwork.

Evaluating worker performance. You must appraise and evaluate the performance of your crew to make sure that they are doing their jobs right. This can only be done by actually watching them as they work. Sometimes testing may be in order. Evaluating a worker's performance allows the manager to see how effective training has been, to assess the ability of the worker to follow instructions, to see what skills the worker has learned, to determine whether the worker accepts responsibility, and to see how well the worker handles stressful or emergency situations.

Summary

Why do we need management? The purpose of management is to balance the needs of the five interests in any business: the investor, the worker, the consumer, the competitor, and the government. *Who is management?* Management includes anybody in a business who is responsible for the work of others. *What does management do?* Managers plan, instruct, and coordinate.

Lesson 2

WHAT IS LEADERSHIP?

OBJECTIVES

Upon completing this lesson, the student will be able to

- *Identify the six basic "wants."*
- *Discuss ways that a manager can help workers to feel more secure in their jobs.*
- *Describe ways to give a worker recognition on the job.*
- *Discuss how to encourage people to gain a sense of accomplishment from their work.*
- *Explain how concern for the family can affect worker performance.*
- *Describe ways to give workers change or diversion in their jobs.*

Compared to a human being, a drilling rig is a simple mechanism. An instruction manual usually comes with new equipment. No such book accompanies a new employee. This lesson focuses on leadership, that is, the ability to motivate people to do what you want them to do. For a supervisor, that means the ability to create a work atmosphere in which people will become motivated to do what needs to be done. You must provide leadership to be able to coordinate work—one of the three basic management skills discussed in Lesson 1. Coordinating work means getting employees to cooperate with each other and to do their work as planned. To accomplish this, they must be motivated to do what you want them to do.

The key to leadership is knowing what people want. That's because people will generally do what you want them to do if they believe that it will help them to achieve something that *they* want. The challenge to a manager is to convince people that working will help them to get what they want. To be an effective manager, then, you must understand what people want and, in particular, what they want so strongly that they are motivated to work for it.

WHAT DO PEOPLE WANT?

Leadership involves helping people to achieve the six basic "wants"

Six basic things that nearly all people strongly desire have been identified. They are (1) food, clothing, and shelter; (2) security; (3) recognition; (4) sense of accomplishment; (5) benefits for the family; and (6) change or diversion. By helping your workers to satisfy some of these basic "wants" through their jobs, they become motivated to do what you want them to do.

Food, clothing, and shelter

The necessities of life—food, clothing, and a decent place to live—are the strongest of the six basic wants. Most people get and keep a job primarily to earn money so that they can buy these basic necessities. But people want more than just the bare minimum to get by—they want to maintain a standard of living similar to that of others around them. Although you may not be directly involved in helping workers to satisfy their desire for a comfortable standard of living, you should always be aware that this desire is the primary reason why most people work.

Food, clothing, and shelter are the strongest of the six basic "wants" for most people.

Security

Security is high on the list of basic wants for many people. A person acquires security, or peace of mind, by being confident that he can provide for himself and his family and that he need not fear the years to come. Workers like to feel that they can count on being employed for the foreseeable future. They also like to know where they stand with their manager and with the company. On the job, those that strongly desire security are likely to work hard and stay with one company for a long time.

Security comes primarily from self-confidence. Some of the ways that you can help your workers to feel more secure on the job include

- *Train workers well.* Lack of knowledge about his job makes a worker feel insecure. The more he knows about his job, the less he'll need to fear losing it.
- *Let each worker know how he is getting along.* It's hard for a worker to build self-confidence if he's always kept guessing. Let him know his strong and weak points so he will know where to improve.
- *Promote on-the-job safety.* By teaching and following safe practices, you can cut down on the number of job-related accidents. Safety increases a worker's security about his physical well-being.
- *Discuss company strengths and future plans.* Confidence in the future of the company will increase a worker's confidence in his job and, consequently, his sense of security.

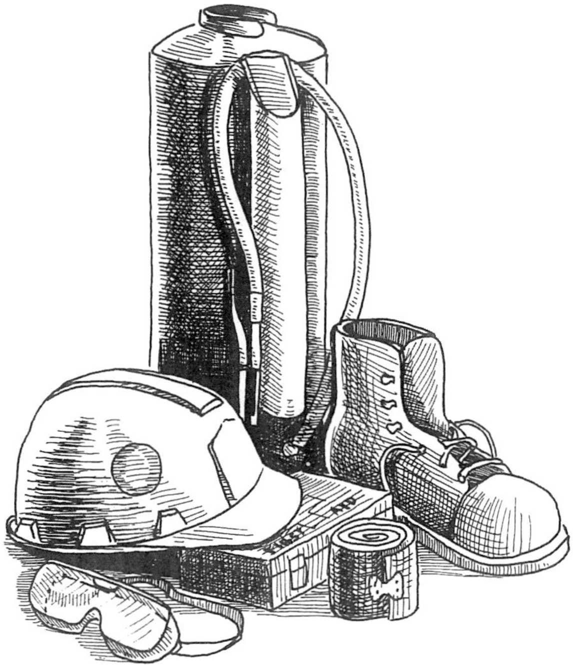

Promoting safe work practices will increase an employee's sense of security on the job.

- *Discuss company benefits programs.* Insurance, retirement, and other employee benefits programs are really security programs.
- *Keep workers informed about how the company is doing.* The more that people know about their job and the company, the more secure they will feel about them. The unknown and imagined are the greatest source of fear.
- *Tell people in advance about changes that will affect them.* People will accept changes more positively if given time to prepare for them.

Recognition

The third basic want is recognition—feeling that you are important and that you are accepted by other people. For some people, the desire for recognition is very strong. If they don't get recognition in positive ways, then they tend to seek it negatively. For example, people who repeatedly take foolish chances on the job, spend beyond their means, or brag about who or what they know are usually craving recognition. Some other ways that people seek recognition are by joining clubs, telling stories or jokes, or running for political office.

You can help workers to satisfy their desire for recognition in positive ways on the job. One sure way is by giving credit where credit is due. If a worker performs well, does extra work, or achieves a higher level of skill at his job, compliment him and reward him in whatever ways are appropriate.

Another way to give workers recognition on the job is to show them that you respect their opinions. Ask your workers for their opinions on how to improve the way things are done, and then listen to their answers. Workers also will gain a feeling of recognition if they are given more responsibility. If a worker knows how to do a job, let him use his own ideas to figure out the details. One word of caution: Failure is recognition in reverse, so don't give a worker more responsibility than he can handle.

Sense of accomplishment

Another basic desire for most people is a sense of accomplishment—a feeling of satisfaction that what they are doing is worthwhile and a sense of pride in a job well done. People for whom this is important tend to take care of their property, keep busy on the job, talk about their work often, and want enough time to do a job well.

You can help to encourage the feeling of pride and satisfaction on the job by showing appreciation for quality work, by setting standards for work high but within reach, and by encouraging additional training for your employees. Another way to help workers gain a sense of accomplishment from their jobs is to encourage them to talk about their work.

Someone who is developing greater interest in his work likes to discuss his achievements and ways for improving his performance. The more interested a worker becomes in his job, the more likely he will feel a sense of accomplishment from it.

Benefits for the family

A fifth want that is important to many people, and is a main reason that they work, is to provide for their family. People want many of the same things for their family that they do for themselves, and more. A worker's concern for his family influences his performance on the job. For example, family men tend to be steady workers who stay at a job longer than their single counterparts. But family problems such as debts or sickness can cause worry and inattention on the job.

To help maximize the positive influence of a worker's concern for family on his work performance, you should take the time to learn more about the employee and his family. Be supportive during times of family sickness or misfortune, and discuss company benefits for the family such as insurance and other programs.

Change or diversion

Change or diversion is the sixth basic want that almost all people desire. Everyone needs a change in the routine now and then to avoid becoming bored. This is especially true on the job, where boredom can lead to carelessness, accidents, inattention to duties, absenteeism, and a high turnover rate.

You can help to relieve boredom or dissatisfaction on the job by rotating work so that no one has to do exactly the same thing every day and by encouraging workers to get more training so that they can take on new responsibilities. You should also encourage your workers to suggest changes that they think would improve the way things are done.

A main reason that many people work is to provide for their family.

HOW CAN YOU USE THE BASIC "WANTS" TO MOTIVATE WORKERS?

An on-the-job situation shows leadership in action

To understand better how you can use the six basic wants to motivate your workers, let's look at an on-the-job situation that demonstrates leadership in action. This situation involves a driller as manager working with one of his crewmen. The same techniques could be used by any other manager.

A leadership scenario

Bobby, the derrickman, is weighing mud in the mudhouse, one of his regular duties. Tom, the driller, stops by to ask how the mud is doing. Bobby tells him that the solids content is high. Bobby then says that he thinks hooking up the desander in a different way might improve the situation. He explains his plan to Tom, who likes the idea and asks Bobby to go with him to ask the toolpusher what he thinks. Tom asks Bobby to explain his idea to the toolpusher, who also likes the plan and tells Tom to go ahead with it. Tom puts Bobby in charge of changing over the desander. Later, Tom offers to get Bobby enrolled in a solids control school that the mud company is offering. Bobby enthusiastically accepts. Tom then tells Bobby that he's pleased with how he's doing his job and glad that he's taken an interest in learning more about mud.

1

Tom listens attentively as Bobby explains his plan for lowering solids content.

2

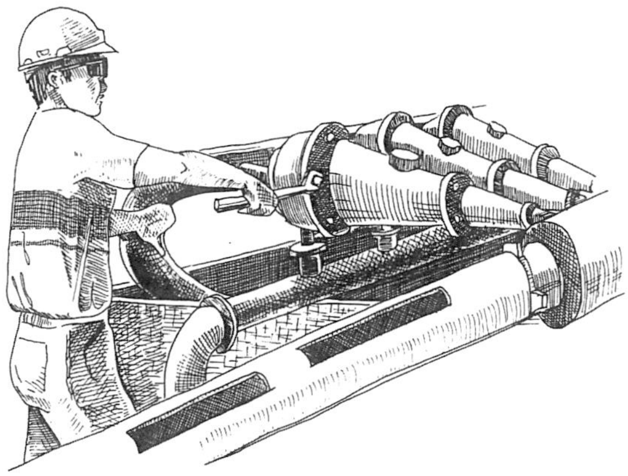

Bobby follows through on his plan by hooking up the desander in a different way.

3

Tom offers to enroll Bobby in mud school and congratulates him for a job well done.

In this scenario, Tom took advantage of several opportunities to help Bobby satisfy basic needs on the job, and thereby demonstrated good leadership. First, Tom gave Bobby recognition by listening to his idea and letting him explain it personally to the toolpusher. Putting Bobby in charge of making the change also showed that Tom has confidence in Bobby, again appealing to his sense of recognition. And Bobby also probably took pride in being allowed to follow through on a plan he had suggested, leading to a greater sense of accomplishment from his job.

By offering to enroll Bobby in the mud school, Tom gave Bobby the chance for a break in his routine, helping to meet his need for diversion or change. Telling Bobby that he's happy with how he is doing probably gave Bobby a greater feeling of security about his job.

By appealing to four of the six basic wants—recognition, sense of accomplishment, diversion or change, and security—Tom helped Bobby to get more satisfaction and pride out of his job. A satisfied worker will do his job well and stay with the company, two definite benefits for the manager.

Summary

What do people want? Nearly all people have six basic "wants": (1) food, clothing, and shelter; (2) recognition; (3) security; (4) sense of accomplishment; (5) benefits for the family; and (6) change or diversion. A leader motivates people to do what he wants by convincing them that so doing will help them to achieve these desires. *How can you use the basic wants to motivate workers?* Giving workers recognition, showing confidence in them, giving them the chance for occasional breaks in the routine, and complimenting them for a job well done are a few of the ways that you can help workers to satisfy some of their basic wants on the job.

Lesson 3

HOW DO YOU HANDLE PERSONNEL PROBLEMS?

OBJECTIVES

Upon completing this lesson, the student will be able to

- ☑ *Describe the four ways that personnel problems are detected.*
- ☑ *Explain the five-step method of solving personnel problems.*
- ☑ *Discuss guidelines to help a manager get information about personnel matters from his workers.*
- ☑ *Show how to fill out a worksheet using the five-step method.*
- ☑ *Discuss the four foundations of good relations.*

Personnel problems are a fact of life in the workplace. Conflicts inevitably arise between people who work together day in and day out. A personnel problem can be defined as any dispute or negative situation involving people where the manager must make a decision and take action. Although even the best manager can't prevent all personnel problems, he should be prepared to quickly and fairly resolve the ones that do arise. This lesson looks at how personnel problems are detected, how to resolve a personnel problem once it has arisen, and how to prevent some future problems.

You usually find out about a personnel problem in one of four ways: the problem comes to you, you sense that the problem exists, you anticipate the problem before it happens, or the problem blows up in your face. A five-step method has been devised for resolving personnel problems. The steps are (1) determine the objectives, (2) get the facts, (3) weigh and decide, (4) take action, and (5) check results. You can use this method to help resolve almost any type of personnel problem.

By following the four foundations of good relations, you can help to prevent some personnel problems. The four foundations are (1) Let each worker know how he is getting along. (2) Give credit where credit is due. (3) Tell people in advance about changes that will affect them. (4) Make the best use of each person's abilities.

HOW DO YOU RECOGNIZE A PERSONNEL PROBLEM?

Most problems are detected in one of four ways

You find out about personnel problems in many different ways. The most common way is that the problem comes to you. Other ways are that you sense that a problem exists, that you anticipate the problem, or that the problem blows up in your face.

When a fistfight breaks out on the rig, the problem has blown up in the manager's face.

The problem comes to you

A crew member approaches you and tells you that he's having trouble getting along with one of the other workers, or that he's dissatisfied with his job, or some other type of problem. That's the most common way that you learn about a personnel problem—it comes to you.

You sense the problem

Production has started to lag, workers who usually are easy-going have been grumbling and arguing lately, or there's a general "down" or depressed attitude on the rig. By picking up on signs such as these, you can sense that a personnel problem likely exists. In this case, you know that you have a problem, but must find out what the problem is.

You anticipate the problem

You are about to make a change in operations that will affect your workers, and you know that some people will have a hard time accepting the change. In this case, you can anticipate a personnel problem before it arises. You must then decide when the best time will be to handle the problem—now or later.

The problem blows up in your face

Two members of the crew get into a fistfight on the rig, or a good worker quits his job in a huff over a minor disagreement. Both of these are personnel problems that blew up in your face, seemingly without warning. This is the worst way that you can find out about a personnel problem, that is, after the problem has already become a major incident. The surest way for a manager to avoid such incidents is to learn to recognize the signs that a problem is brewing before it has a chance to blow up.

HOW DO YOU RESOLVE A PERSONNEL PROBLEM?

The five-step method

We've defined a personnel problem as any dispute or negative situation involving workers where you must make a decision and take action. Some personnel problems are fairly minor, and a brief talk between you and the workers involved may be all that's needed to resolve them. Others are severe and may require more drastic action. The five-step method described below can be used to resolve any kind of personnel problem.

The five steps in this method of problem solving are (1) determine the objectives, (2) get the facts, (3) weigh and decide, (4) take action, and (5) check results. By following each of these steps, you should be able to reach a fair resolution for any personnel problem.

The first step in solving any problem is to determine the objectives, that is, to pinpoint what will be accomplished by solving the problem. Say, for example, that the crew has been arguing a lot among themselves lately. The primary objective would be to get the crew to stop arguing with each other. However, there usually is more than one objective in solving any problem. In this case, a secondary objective might be to decrease personnel turnover. A long-term objective would be to regain a teamwork attitude among the crew members.

Get the facts

Once the objectives have been determined, the next step in problem solving is to get the facts. You need to find out the whole story behind a personnel problem before you can make a decision. First of all you'll need to find out if the situation has been going on for a long time or if there has been a recent change in attitude or behavior. You'll also need to find out what rules or policies apply to the situation, either by talking with someone in the personnel office or by checking the company handbook.

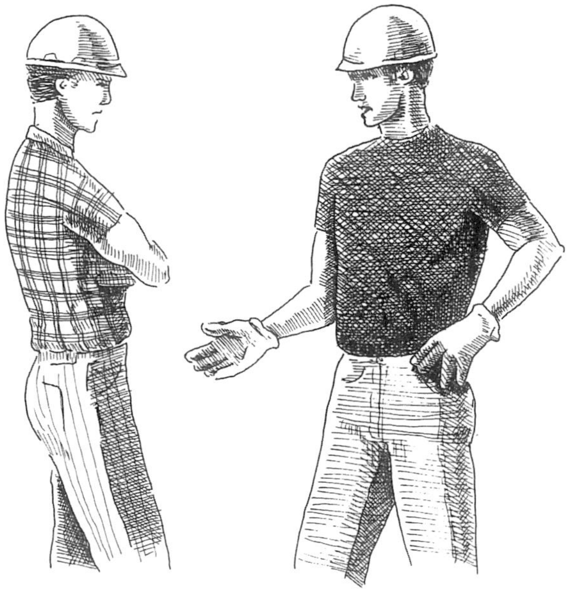

Listen carefully when an employee tells you his side of the story.

Getting the facts will require talking to the individuals concerned, including the persons directly involved with the problem and others who work with them. You must take care to separate facts from opinions, and to treat each accordingly in assessing the situation. The following guidelines will help you to get information from your workers.

Listen. Keep your mind open so that you are better able to understand what the person is trying to say. Don't interrupt when the employee begins talking and let him say anything that he wants to say.

Encourage the person to talk about what is important to him. You may have the tendency to talk about what you think is important. But the goal is to find out what is important to the employee.

Don't argue. Arguing forces a person to defend a position and makes him less receptive to change. If you avoid taking a stand yourself, that allows the employee to remain flexible so that he can change his position without embarrassment.

Don't jump to conclusions. Refrain from making any judgments until the employee has finished talking. Information obtained in the last few minutes can prove an earlier conclusion wrong.

Weigh and decide

Once you have all the facts about a situation, you must then put the facts together and figure out what to do. Consider all the possible actions that could be taken, and try to anticipate the effect that each would have on the individuals involved, on the group, and on production. Take into account the policies and practices of your company.

Take action

After carefully weighing the facts and reaching a decision, you then must take action on that decision. You may decide to refer the problem to your supervisor or to handle the problem yourself. Timing and circumstances of any disciplinary action are important. Reprimands should always be done in private, never in front of other crew members.

Check results

The final step in this problem-solving method is to check results. Decide in advance how and when you will follow up on your actions to make sure that the personnel problem has been solved. You will need to watch for changes in attitudes and relationships among the crew members, as well as keep an eye on the bottom line, namely, whether or not your action improved production.

HOW DOES THE FIVE-STEP METHOD WORK?

Two case studies demonstrate problem solving

In using the five-step method to solve a personnel problem, it usually is helpful to write each step down. You can use the work sheet that we've used for the two case studies on pages 20 and 24 or make up a form of your own. (A blank work sheet for you to photocopy is provided at the end of the lesson.) Writing the steps down allows you to compare all the facts and possible solutions side-by-side, and also lets you make sure that you haven't omitted any important considerations.

To demonstrate how the five-step method works, let's look at two situations involving personnel problems. As you read each case, put yourself in the position of the supervisor. Assume that this is your problem and that you're going to have to do something about it. After each case, we'll show how you might fill out the work sheet as you use the five-step method to solve the problem.

Case 1: A major mismatch

Larry is a unit operator on a well-servicing unit and is responsible for supervising a three-man crew. Larry is supervised by Hal, the foreman. One evening, Larry telephoned the unit superintendent, who is over Hal, to complain that Hal had loudly criticized him in front of his crew that day. Larry told the superintendent that Hal had a grudge against him because four years ago, Hal had been involved in an automobile accident with some of Larry's relatives. The relatives sued Hal and his driver's license was taken away.

Larry is twenty-five years old and has been with the company for four years. He has been considered an average employee since he started with the company and always gets his work done. The main complaint against him is that he is reckless and abuses the company equipment. Larry had been under another foreman before Hal for two years, and that foreman had asked that Larry be transferred.

Larry is married and has two children. He drives a sports car and is reckless with it, just as he is with the company equipment. He has been given several warnings for reckless driving by the highway patrol. Larry was raised in a large city. The members of his crew grew up in the country and live in rural districts now. They hunt and fish together. Larry does not mix with the men socially.

Hal is forty-five years old and has been with the company for two years. He has twenty-five years of experience in the oil patch and is considered to be the best foreman in the district.

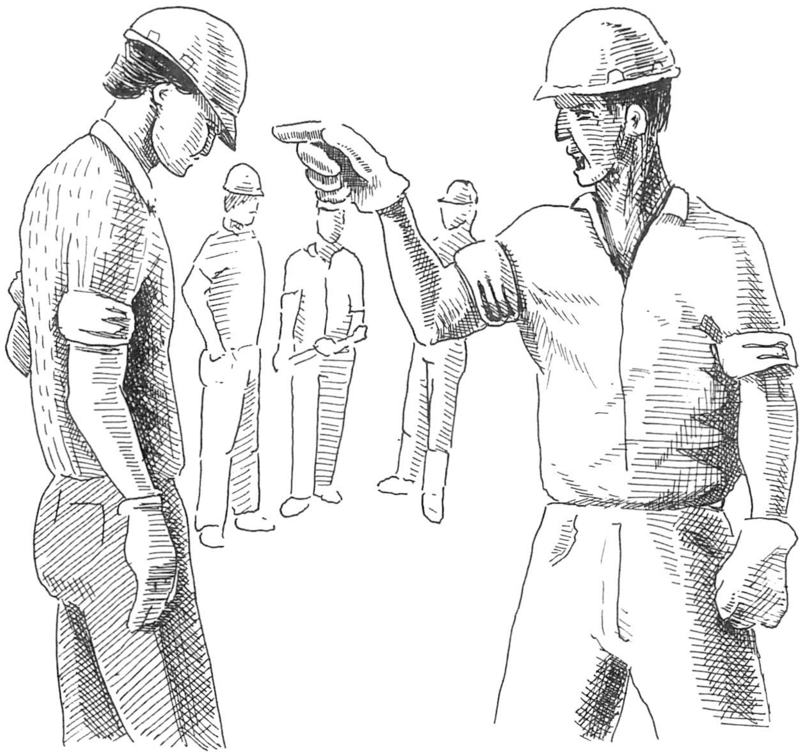

Hal criticizes Larry in front of his men.

PROBLEM-SOLVING WORKSHEET

1. Determine the objectives
2. Get the facts
3. Weigh and decide
4. Take action
5. Check results

Objectives:

1. Get crew working as a team 2. Change Larry's behavior 3. Increase work output

Facts:

1. Larry said Hal criticized him in front of men 2. Larry said Hal had grudge against him 3. Hal involved in accident with Larry's relatives 4. Larry has been an average employee 5. Larry is reckless and abuses equipment 6. Hal has 25 years experience in oil field 7. Hal is considered best foreman in district 8. Larry does not mix socially with his men

Possible actions:

1. Fire Larry 2. Fire Hal 3. Talk to Larry about his work record (prefer that Hal do it) 4. Talk to Hal about criticizing Larry in front of men 5. Demote Larry 6. Transfer Larry to another unit 7. Get Hal and Larry talking

Action:

combine 3, 4, and 7

Follow-up: 1. Visit rig first week after Larry and Hal talk
2. Talk to Hal and see how it went
3. Enroll Hal & Larry in supervisory training course

Put yourself in the position of the unit superintendent who got the call from Larry, and use the five-step method to try to figure out what to do about the situation.

Determine the objectives. This personnel problem is really several different problems. It will therefore involve several objectives. The most obvious would be to get Hal, Larry, and the crew working together as a team. A second would be to change Larry's reckless behavior. A long-term goal would be to increase the crew's output.

Get the facts. Many facts and opinions are involved in this case. Some of the important ones: Hal criticized Larry in front of his men. Larry said that Hal had a grudge against him because of the accident involving his family. Larry has been an average employee but is reckless and abuses equipment. Hal has twenty-five years of experience in the oilfield and is considered to be the best foreman in the district.

As in any case involving personnel problems, all the information should be verified before it is accepted as fact. The superintendent should check with crew members who witnessed the incident, for example, and look at the personnel records of both Larry and Hal.

Weigh and decide. Several things are apparent from the facts of this situation. First, steps should have been taken before the situation got this bad. Larry's previous foreman should have talked to him about his reckless behavior. And Hal should have told someone about his trouble with Larry's relatives or spoken to Larry when he was first transferred. Larry is mismatched with Hal and with his crew. The personnel office or whoever was responsible for Larry's transfer should have foreseen the trouble.

What are the superintendent's options in this situation? He could fire Larry or fire Hal, or both. He could talk to Larry about his work record, or ask Hal to talk to him about it. He also could talk to Hal about criticizing Larry in front of his crew. He might consider demoting Larry, or transferring Larry to another unit. Finally, he could try to get Larry and Hal to talk with each other and work out their differences.

Take action. In this case, the best solution seems to be a combination of some of the above options. The superintendent should talk to Hal about criticizing Larry in front of his crew, and then ask Hal to talk to Larry in private about his reckless behavior. The superintendent also should try to get Larry and Hal to work things out betweeen themselves.

Check results. Because the situation is so volatile, the superintendent must be certain to follow up on his actions. If Hal and Larry aren't able to work things out between themselves, then further action may be necessary. He should plan to visit the unit shortly after talking with Hal to see if the situation has improved. He also might consider enrolling both Hal and Larry in supervisory training courses—Hal to learn the right way to reprimand workers, and Larry to learn how to get along better with his crew.

Hal talks to Larry in private about reckless behavior.

Case 2: A managerial mess

Jimmy is a driller on the morning tour at Godeep Drilling Company. His rig is about 60 miles from his home. All of his crew meet him at his house each morning and he drives them to the rig in his car. He likes this arrangement because it gives him a chance to get to know his crew.

Until recently, Jimmy's crew had been doing a good job. In fact, his crew had been the fastest, safest, and cleanest crew in the company. But while drilling the last two wells, the derrickman missed several stands, the floorhands had been arguing a lot, and one engine threw a rod because it was low on oil. Lately, Jimmy seemed to be spending most of his time telling the crew members to watch themselves or reminding them what their jobs were.

Perplexed, Jimmy went to his toolpusher, Punk, to ask him what he thought might be bugging his crew. Jimmy told Punk that he had overheard his crew complaining that other companies had better benefits than Godeep. He asked Punk if he thought that this might be the reason for their bad attitude.

Jimmy's crew stands around arguing—something that they've been doing a lot of lately.

Punk told Jimmy that he thought his crew was a bunch of "whiners," and that they had a lot more benefits than he had in the "old days." He told Jimmy that his "long-haired derrickman doesn't care about anything except getting his paycheck and smoking dope." Jimmy responded, "He does have long hair, but I have never known him to use drugs and never heard that he did."

Punk then told Jimmy that his people think they deserve special treatment just because they trip in and out at an acceptable speed. "The other day, I caught them all standing around the mud pumps doing nothing," Punk continued. "I told them to get to work or that I would find something for them to do so that they could use their so-called talents. I told them that just because they were sort of fast, that didn't mean that they could loaf. When I was a floorhand, you had to prove yourself every day." Punk told Jimmy that in his thirty years in the oil patch, he had never seen a bunch of young people who cared as little as the ones around nowadays.

Jimmy was only twenty-eight years old himself, with eight years in the oil patch, and Punk's attitude frustrated him. Jimmy thought to himself that maybe Punk was right, and that he should fire his whole crew and get another one. After all, he thought, Punk always treated him alright and he sure knew downhole frontward and backward.

Put yourself in the position of Jimmy, the driller, and use the five-step method to try to decide what to do about the situation.

Determine the objectives. Jimmy's first objective would be to get his crew back on track, working well together as they had before. Other objectives would be to improve the crew's attitudes and also to improve production. But another objective appears warranted, namely, to improve the toolpusher's attitude. Punk's negative attitude toward the crew seems to be a big part of the problem. So to get the crew thinking positively again, Jimmy must also do something about Punk.

Get the facts. Among the facts in this situation: The crew used to be a good crew. Recently, the derrickman missed stands, the floorhands have been arguing, and an engine threw a rod because it was allowed to get low on oil. Jimmy also has had to spend a lot of time lately correcting the crew. Jimmy overheard the crew complaining about company benefits. Punk said that the crew was a bunch of "whiners." Punk accused one crew member of smoking marijuana. Punk told Jimmy that he had criticized the crew recently for standing around doing nothing. Punk has been in the oil patch for thirty years.. He has always treated Jimmy well and knows downhole operations very well.

Weigh and decide. In thinking over the facts of the situation, Jimmy realized that his crew had never received much positive recognition when it performed well. Punk had given them recognition, alright, but a very negative and destructive kind. Punk is the real problem in this situation. He went around his subordinate supervisor, Jimmy, to chew out his crew. He even seemed to resent the crew for its accomplishments.

What can Jimmy do to resolve this situation? He could fire his crew and get a new one, or he could quit his job and avoid the whole mess. He could ask Punk to arrange for the office staff to talk to his crew about company benefits. He also might try to give

PROBLEM-SOLVING WORKSHEET

Objectives:

1. get crew back on track 2. improve crew's attitude 3. improve Punk's attitude 4. Increase production

1. Determine the objectives
2. Get the facts
3. Weigh and decide
4. Take action
5. Check results

Facts:

1. Crew used to be good 2. Derrickman missed stands 3. Floorhands arguing 4. Oil in engine low--threw rod 5. Lately crew needed lots of correcting 6. Crew not happy with benefits 7. Asked Punk for help 8. Punk said crew is a bunch of "whiners" 9. Punk accused one crew member of smoking marijuana 10. Punk accused crew of not caring about anything 11. Punk chewed out crew for standing around doing nothing 12. Punk in oil patch for 30 years
13. Punk knows downhole operations well

Possible actions:

1. fire crew 2. quit job 3. ask Punk to get office staff to talk about benefits with crew 4. give crew positive reinforcement as well as negative 5. talk to drilling superintendent about Punk 6. ask Punk to talk to crew about downhole 7. get the crew's record published in company magazine

Action:

combine 3, 4, 6, and 7

Follow-up: 1. next crew change, follow up with Punk on benefits
2. get article written for Punk to give to editor within a week
3. survey attitude of crew and Punk after downhole talk
4. keep record of good & bad situations to discuss with crew, and don't harp on the negative (ongoing)

the crew more positive reinforcement. Another option would be to report Punk's behavior to the drilling superintendent. He could ask Punk to have a talk with his crew about downhole operations. He also might try to get an article about the crew's superior work record in the company magazine.

Take action. For Jimmy, the best solution seems to be a combination of these options: Ask Punk to have benefits explained to his crew, and also ask Punk to talk with the crew about downhole operations. He also should make an effort to give the crew more positive reinforcement, such as the write-up in the company magazine.

Check results. Jimmy will need to follow up on his actions to make sure that the company benefits are explained to his crew and that Punk had the chance to talk with them about downhole operations. He also must see to it that the article for the company magazine is written and submitted to the editor. He will need to survey the attitudes of the crew and of Punk after the downhole meeting to see if things have improved. He also might want to keep a record from then on of good and bad situations and discuss them with the crew periodically, making sure to stress the positive.

A write-up in the company magazine gives a crewman positive recognition.

HOW DO YOU PREVENT PERSONNEL PROBLEMS?

The four foundations of good relations

There's no way that you can prevent personnel problems from ever happening. People are people, with different personalities and moods, and nobody can make them get along with each other perfectly and do their jobs perfectly at all times. You can, however, take certain measures to make sure that *you* are not the direct cause of a personnel problem.

Lesson 2 on leadership discussed how you can help workers to meet basic needs on the job. Four rules, called the foundations of good relations, have been derived from those basic wants. By abiding by the four foundations of good relations, you can help your workers to get greater satisfaction out of their jobs and to get along better with each other. Satisfied workers mean fewer personnel problems.

Let each worker know how he is getting along

If Larry, the reckless unit operator in our first case, had been told by his original foreman that his abuse of the equipment could not be tolerated, Larry might have shaped up sooner. The blowup with Hal might never have happened. Everyone likes to know where they stand with the boss and with the company. Let your workers know exactly what's expected of them and then how well, in your eyes, they are meeting those expectations. Find ways to give new employees recognition while they are learning and point out ways that they can improve.

Give credit where credit is due

Always be on the lookout for extra effort or superior performance and find ways to reward such behavior. If Jimmy, the driller in our second case, had given his crew more positive feedback when they were doing well, then Punk's outburst might not have bothered them so much.

Tell people in advance about changes that will affect them

Abrupt changes in the way things are done, dictated from above, are among the most frequent sources of personnel problems. Changes handled in this way can make a worker feel helpless and frustrated, and very dissatisfied with his job. By warning workers in advance of changes that are coming, and if possible explaining why changes are necessary, you can help to avert worker dissatisfaction.

Make the best use of each person's abilities

If a worker feels that his talents are being wasted in his job, he will likely be dissatisfied with it and not do it very well. Always look for ability not being used, and find ways to put it to work. Likewise, encourage your workers to advance themselves in the company.

Summary

How do you recognize a personnel problem? Personnel problems are detected when a problem comes to you, when you sense that a problem exists, when you anticipate that a problem will occur, or when a problem blows up in your face. *How do you resolve a personnel problem?* The five-step method of problem solving is (1) determine the objectives, (2) get the facts, (3) weigh and decide, (4) take action, and (5) check results. *How do you prevent personnel problems?* You can help to avert personnel problems by following the four foundations of good relations: (1) Let each worker know how he is getting along. (2) Give credit where credit is due. (3) Tell people in advance about changes that will affect them. (4) Make the best use of each person's abilities.

PROBLEM-SOLVING WORKSHEET

Objectives:

1. Determine the objectives
2. Get the facts
3. Weigh and decide
4. Take action
5. Check results

Facts:

Possible actions:

Action:

Follow-up:

Lesson 4

HOW DO YOU START OUT A NEW HAND?

OBJECTIVES

Upon completing this lesson, the student will be able to

- ☑ *Define the term* morale *as it relates to workers on the job.*
- ☑ *Discuss the relationship between worker morale and productivity.*
- ☑ *Describe the three-part program for orienting new employees.*
- ☑ *Discuss some of the topics that a supervisor should bring up during the initial interview with a new employee.*
- ☑ *Explain the importance of safety training in an orientation program.*

On the morning that a new employee reports on the job, his morale level is 100 percent. He wants to do a good day's work and prove that somebody made the right decision when they hired him. He will watch every move, listen to every word, and try to get along. Although his morale is high, however, it is very unstable. A new employee is strongly influenced by the attitudes of his coworkers. To be accepted and fit in, he will adjust himself to conform to their feelings, attitudes, and loyalties toward their jobs.

This lesson looks at the right way to orient a new worker—that is, what to do in those critical first few days to get the new worker off to a good start. Morale, or an employee's attitude toward his job, coworkers, and the company, is directly related to productivity. The higher the morale, the higher the productivity. If morale drops below a certain point, the worker will quit. So to have and keep productive employees, you must keep morale high.

A good orientation program helps to maintain the high morale of a new worker. Most orientation programs consist of three parts: initial interview, introduction to the job, and follow-up. If you follow through on each of these parts, chances are good that the new employee will develop a positive attitude toward his job and become and remain a productive worker.

WHY IS ORIENTATION IMPORTANT?

Getting workers off to a good start

A worker's morale might be defined as the attitude that he has toward his job, his coworkers, his opportunities for advancement, his economic situation, and his physical well-being on the job. If a worker does not receive a good orientation, his morale can drop sharply during the first few days of employment. Once the employee becomes adjusted to the organization, morale will stabilize and may be hard to change. That's why an employee's first impressions are so important. The opinions he forms during the first few days on the job can be lasting ones.

The situation described in the box below illustrates what can happen to the morale of a worker who is not oriented on his first day on the job.

Mike started out on his new job with high morale. But as the day progressed, his morale began to drop. And if things continue as they are going over the next several days, Mike may be looking for a new job by the end of the week.

Charting worker morale (and productivity)

Think of charting worker morale on a graph, with the base line representing time and the vertical line representing morale. In Mike's case, the graph would take a very sharp plunge after only one day. Unless things change drastically, it's unlikely that the curve will ever swing upward.

With no orientation, Mike was bewildered his first day on the job.

What *not* to do on a new hire's first day

Mike was not sure where or what the stack was, but he did not wait around. He changed into his work clothes and began looking for the other workers. As the day progressed, Mike tried hard to keep up. He was never really sure that he was doing things correctly, but he was afraid to ask too many questions. He didn't want to look stupid.

Mike had time to think about the day's activities while driving home. He liked the hard work and good pay, but felt like a stranger. He wanted to do things right but was unsure of safety rules and company policies. Mike just didn't know what to think about his new job.

Mike was hired by the superintendent of a drilling company to work as a floorhand. He was

On the same graph, picture a "quit line" representing the level at which the worker's morale has dropped so low that he will quit and look for a better job. The quit line is floating; it rises when jobs are plentiful and falls when they become difficult to find. The quit line also falls as a man gets older and finds it more difficult to change jobs. Because Mike did not receive proper orientation, his morale was very close to the quit line after only one day on the job.

Apart from the possibility that low morale will force a worker to quit his job, a second reason for keeping morale high is productivity. The relationship between morale and productivity is direct. That is, the higher a worker's morale, the greater his productivity; the lower the morale, the lower productivity. On the graph plotting worker morale, the curve labeled *morale* might instead be labeled *productivity.* An enthusiastic worker who is happy with his job will produce more than a dissatisfied worker who is always on the verge of quitting his job.

not sure of the job responsibilities but was confident that he was capable of doing the work. Many of his friends had worked in the oil patch. After taking a physical, Mike was told to report to his supervisor, the driller, at 6 o'clock the next morning.

Although he left the house early, Mike didn't find the location until 6:30. He was amazed by the large equipment and all the noise. It was not what he had expected. After asking several people, Mike located the driller.

The driller briefly introduced himself and then began lecturing Mike about reporting to work on time. He ended the conversation quickly, telling Mike to help the rig crew nipple up the stack and not to get caught without his hard hat.

Worker morale

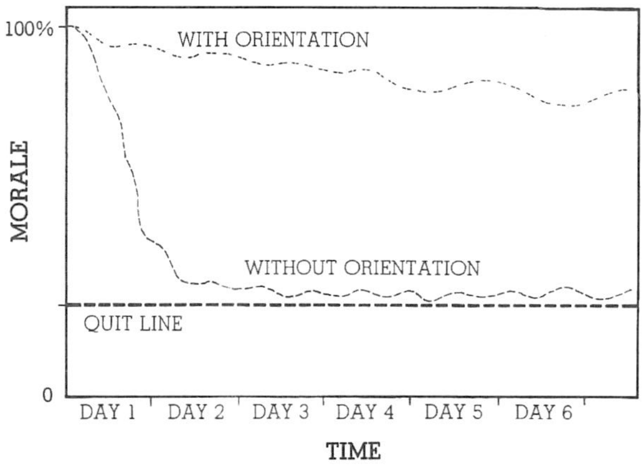

With orientation, worker morale stays high. Without, morale hovers just above the quit line.

WHAT IS THE ORIENTATION PROGRAM?

A three-part plan for bringing new people onboard

An orientation program provides an opportunity for a new employee to learn about the company and how things work in general. He also will learn exactly what his job is and how he fits into the organization.

Most of the orientation should be given by the new employee's immediate supervisor. In fact, much of it can be given by no one else. This helps to build a stronger relationship between employee and supervisor, encouraging the employee to look to his supervisor as a source of information. Others in the organization, however, can assist in the orientation process. Someone from the personnel office, for example, might be called upon to explain benefit programs to the new employee. And all rig hands should pay special attention to the new employee to make sure that work continues safely.

The orientation program should be done under the assumption that the immediate supervisor was not present during the interview with the new employee before he was hired. The supervisor should question the employee about his expectations for the job. That way, if there are any misunderstandings, they can be cleared up early.

One goal of the orientation program is to get a new employee started out right while morale is high and he is anxious to do a good job. Many things happen during the first few days on a new job, and the employee should not be expected to remember everything. Information should be given and repeated over time until it is thoroughly covered.

A planned orientation program is usually recommended; however, this may not be appropriate for every situation. One company has a form that lists all items in the orientation program. Each item is checked off and dated by the supervisor as it is done.

Orientation should be given during the first day or first few days of employment. The length of time given to the orientation program will vary, depending on the nature of the business and the job of the new employee. Conditions on the rig and the necessity of keeping the rig operating are important considerations. Exact content of the orientation also will vary, but the program generally consists of three parts: initial interview, introduction to the job, and follow-up.

Initial interview

As the first step in the orientation program, you should welcome the new employee aboard. This would include a brief visit to get acquainted, when you should try to make the new employee feel relaxed. You might try to establish a mutual interest, or inquire about the new employee's family. Other subjects to be discussed during the initial interview include

- the company's aims, philosophy, policies, and organizational structure;
- the organization of the rig and its relationship to other departments or units in the company;
- the employee's job and how he fits into the organization;
- proper clothing and safety equipment;
- local practices, including safety procedures and how to handle tools and equipment.

A tour of the rig helps a new worker feel like he belongs.

A brief chat at the end of the first day allows a new worker to ask any questions he may have.

You should take the new employee on a tour of the operation, pointing out where things are and where he will be working. During the initial interview, you should also tell the new worker when payday is and make sure that he understands the pay. Training programs and long-range opportunities also should be mentioned.

Introduction to the job

After the initial interview, you should explain to the new employee exactly what his duties will be. This includes information on how to check out the tools he will be using and where to obtain supplies or other materials. You should then give the worker a specific task to perform, telling him who to go to with questions if you are not available to answer them. Emphasize that there is no such thing as a dumb question and that asking questions is a good way to learn.

Follow-up

Follow-up is the final step in the orientation program. You should allow some time at the end of the worker's first day on the job to answer any questions he may have. You should then check with the employee to see if he has further questions over the next few days, and be available as needed. Check in with the employee on his first payday to see that he understands the pay. Also set aside a time for follow-up before the end of the probationary period, if there is one.

Orientation and safety

Orientation is the ideal time to set safety standards for a new employee. Throughout orientation, you should stress the importance of using safety equipment and adhering to safe work practices. All new employees should be given an explanation of the company's accident prevention and safety program. Keep in mind that early training in safety will result in a lower accident rate down the line.

A manager should stress safety throughout orientation.

Orientation for the transferred or veteran employee

This lesson has been directed mainly to orientation for new hires from outside the company. Much of the same program would apply to an employee transferring from some other part of the company. He is still a stranger to the new operation, and needs to be accepted by his supervisor and coworkers and become familiar with the way things are done. And although all rigs are basically the same, some equipment and the way it is arranged will be new to the transferred employee. There are also differences between land and offshore rigs and between drilling and workover rigs, for example.

The same applies to an experienced employee who has worked for other contractors but is new to an operation. Just as with the transferred employee, he needs to learn the company's way of doing things and what is expected of him as a crew member. In the case of both the transferred and the veteran employee, you can avoid needless repetition by questioning the new hire during the initial interview about what he already knows.

Summary

Why is orientation important? A worker's morale, and long-term productivity, are often established for good after his first few days on the job. Orientation helps to get a new worker started out right. *What is the orientation program?* An orientation program usually consists of three parts: initial interview, introduction to the job, and follow-up. The importance of safety should be stressed throughout orientation.

REVIEW QUESTIONS

LESSONS IN ROTARY DRILLING

Unit IV: Man Management and Rig Management

Listed below at left are the five interests in any business. At right are the primary concerns of each of these interests. To answer questions 1–5, write each letter in the blank next to the interest with which it corresponds. Use each letter only once.

1. ______	investor	A. price, quality
2. ______	worker	B. wages, secure employment
3. ______	consumer	C. lure away customers
4. ______	competitor	D. taxes, regulation
5. ______	government	E. profit, secure investment

6. ______ When a manager gives orders, his tone of voice usually has little impact on how the orders will be received by his employees. (True or False?)

7. ______ Which of the following is a manager responsible for?
   A. planning the work of employees
   B. training employees
   C. getting employees to work together
   D. all of the above

8. ______ Keeping a worker uninformed about company business increases his sense of security about his job. (True or False?)

9. ______ To help a worker achieve a feeling of pride in his job, you should—
   A. set work standards low so they are easy to reach.
   B. discourage any outside training that might cut into work time.
   C. show your appreciation for quality work.
   D. discourage him from talking much about his job.

10. ______ For most people, which is the strongest of the six basic "wants"?
   A. security
   B. recognition
   C. benefits for the family
   D. food, clothing, and shelter

11. ______ Boredom on the job can lead to carelessness and accidents. (Tue or False?)

12. ______ In what order should the five steps in problem solving be performed?
   A. (1) weigh and decide, (2) get the facts, (3) determine the objectives, (4) take action, (5) check results
   B. (1) determine the objectives, (2) get the facts, (3) weigh and decide, (4) take action, (5) check results
   C. (1) take action, (2) get the facts, (3) weigh and decide, (4) determine the objectives, (5) check results
   D. (1) get the facts, (2) weigh and decide, (3) take action, (4) check results, (5) determine the objectives

13. ______ If somebody repeatedly takes foolish chances on the job, which basic "want" is he likely craving?

A. recognition
B. sense of accomplishment
C. change or diversion
D. security

14. ______ Which of the following is the *worst* way for a manager to find out about a personnel problem?

A. A problem comes to the manager.
B. The manager senses that a problem exists.
C. The manager anticipates that a problem will arise.
D. A problem blows up in the manager's face.

15. ______ When a manager is trying to get the facts about a personnel problem, it is a good idea to—

A. make up his mind about the situation before he talks to the people closely involved with the problem.
B. set firm guidelines as to what employees may talk about when discussing the problem with them.
C. avoid arguing with the employees when discussing the problem with them.
D. do all of the above.

Listed below are some ways that a manager might treat his employees. In the blank next to each number, write A *if the approach would likely* prevent *personnel problems,* B *if it would likely* cause *problems.*

16. ______ Keep a worker guessing about how he is getting along.

17. ______ Find ways to reward extra effort or superior performance.

18. ______ Don't comment on a worker's performance unless a serious problem comes up.

19. ______ Warn people in advance about changes that will affect them.

20. ______ Encourage workers to advance themselves in the company.

21. ______ A manager should try to keep worker morale high so that—

A. workers won't quit their jobs.
B. workers will be enthusiastic.
C. worker productivity will be high.
D. all of the above.

22. ______ Most of a new worker's orientation program should be given by—

A. fellow workers.
B. the personnel office.
C. the immediate supervisor.
D. the company president.

23. ______ Productivity of an enthusiastic worker and of a dissatisfied worker will usually be about the same. (True or False?)

24. ______ During a new employee's first day on the job, it is a good idea for a manager to—

A. give the new employee a job and then let him figure it out on his own.
B. discourage the new employee from asking too many questions until he knows more about how things are done.
C. take the new employee on a tour of the operation.
D. do none of the above.

25. ______ The orientation program should be given no later than the new employee's ______________________ of employment.

A. first few days
B. first few weeks
C. first few months
D. first year

At left below are some statements based on the laws of learning, which are listed at right. To answer questions 26–30, write the letter of the law of learning in the blank next to each statement with which it corresponds. You may use each letter more than once.

26. ______ The more often an instruction is repeated, the more likely it will be learned.

27. ______ A person won't respond until he wants to respond.

28. ______ If a person feels successful after learning a new job, he will be interested in learning other new jobs, too.

29. ______ A child learns multiplication tables by repetition.

30. ______ If a person fails at learning one job, he will avoid trying to learn other jobs.

A. law of readiness
B. law of effect
C. law of exercise

To answer questions 31–35, write the letter of each description on the right in the blank next to the corresponding term on the left. Use each letter only once.

31. ______ key point
32. ______ progress chart
33. ______ job
34. ______ job breakdown
35. ______ step

A. task that can be completed in a short period of time
B. action that causes a job to progress toward completion
C. tip to help worker perform a step
D. keeps track of a worker's training
E. lists steps in a job

36. ______ Which of the following will help a driller to get along better with the next driller coming on tour?

A. passing information on
B. offering to help out
C. keeping the rig clean
D. all of the above

37. ______ To improve communications with your boss, you should keep him informed about any problems and make recommendations to him after thinking through the problems carefully. (True or False?)

38. ______ How can a first-line supervisor help to maintain good working relationships with staff people?

A. by asking for help only when a major problem has arisen
B. by holding off on supplying information to them
C. by asking for advice before trouble arises
D. none of the above

39. ______ On the day that a casing crew comes to the rig, the driller should—

A. have ready all the tools, equipment, and personnel that the crew will need.
B. welcome the crew chief to the rig.
C. let the crew chief know that he can come to him with any questions or problems.
D. do all of the above.

40. ______ An active manager—

A. avoids the risk of being blamed if things don't turn out as expected.
B. lets things happen and handles problems as they come up.
C. makes decisions about how things should be done and bears responsibility for them.
D. does none of the above.

41. ______ Planning can help to—

A. increase worker productivity.
B. cut costs.
C. prevent abuse of equipment.
D. do all of the above.

42. ______ No amount of planning can help prevent accidents on the rig. (True or False?)

43. ______ Flexibility is important in planning—that is, you must be ready to change your plan if circumstances change. (True or False?)

44. ______ A Gantt chart is used to—

A. break down jobs into steps.
B. keep track of a worker's training.
C. schedule jobs.
D. solve personnel problems.

45. ______ Once a manager has written down a job schedule, he should always follow it to the letter. (True or False?)

46. ______ The only way to reduce the costs of your operation is to cut back on spending. (True or False?)

47. ______ To improve profits, you sometimes must increase spending. (True or False?)

48. ______ A manager can improve efficiency on the job by discouraging outside training that would cut into an employee's work hours. (True or False?)

49. ______ The longer the *payout period* on a proposed change, the sooner it will begin to save money for the company. (True or False?)

50. ______ The *period of improved profit* for a proposed change refers to the length of time that a change will continue to save money after the payout period. (True or False?)

Lesson 5

HOW DO YOU TRAIN NEW EMPLOYEES?

OBJECTIVES

Upon completing this lesson, the student will be able to

- ☑ *Identify the three laws of learning.*
- ☑ *Discuss the four guidelines that apply to most training situations.*
- ☑ *Define the term* job *as it applies to a training program.*
- ☑ *Explain how to use an employee progress chart to track a training program.*
- ☑ *Describe how to break a job down into steps.*

Should you train your workers or not? That's a pointless question, because new workers will be trained one way or another. If you don't train them purposely, they'll pick things up haphazardly from other crew members or by trial and error. This kind of training by accident, unfortunately, hardly ever gives the worker what he needs to do the job safely and efficiently. The real question, then, is should you use a planned program to train your new people? The answer is absolutely yes. A planned training program will, in the long run, reduce the time and money required for developing competent rig workers. It will help workers make the best use of their abilities and prepare them for advancement.

This lesson looks first at how people learn. Three laws of learning have been identified: the law of readiness, the law of effect, and the law of exercise. These laws are the basis for four training guidelines that can be used in most situations. You should (1) get the new worker ready to learn, (2) tell and show how the job should be done, (3) let the new worker do the job, and then (4) follow up on the new worker's progress.

This lesson also discusses how to plan a training program. First, you should compile a list of all jobs that the new worker will be expected to perform. A progress chart will help you keep track of the employee's training for these jobs. Breaking down each job into steps and noting key points will also help during training.

HOW DO YOU LEARN?

Stimulus-response is the model for the laws of learning

Before you can train, you have to know something about how people learn. The most basic way that you learn about anything is through your five senses and the body's central nervous system. Sensory nerves in the eyes, ears, nose, mouth, and skin pick up information and send it to the brain, which responds to the information or stores it until it is needed. For example, if a child puts his hand on a hot stove, sensory nerves in his hand send a message to the brain that the stove is hot. The brain then sends a message to his hand to move it off the stove. The action going on here is stimulus and response, that is, the sensory nerve is stimulated and the brain responds. Learning is a complex process involving much more than simple stimulus-response, such as the individual's own abilities and many other variables. The stimulus-response mechanism does, however, serve as a useful model for understanding some of the basic aspects of learning.

Stimulus: The stove is hot. Response: Move hand off!

To look at it another way, think of the mind as a storage tank for everything that you learn. The five senses are the "pipelines" that feed it. Everything that you learn comes to you through one of these pipelines: sight, hearing, smell, taste, and touch. The information that you receive from these pipelines is stored in the brain, so that it can be recalled, or responded to, later.

Because a person can learn through each of the five senses, training should involve as many of them as possible (five pipelines can deliver a lot more oil to a storage tank than one alone). So, for example, you should not just *tell* a new worker how to perform a particular job, you should *show* him how as well. By telling *and* showing, two senses are stimulated—hearing and sight. If you also allow the trainee to physically do the job during training, then a third sense, touch, also comes into play.

The laws of learning

Using the stimulus-response mechanism of the central nervous system as a model, three laws of learning have been identifed. They are the law of readiness, the law of effect, and the law of exercise. To put it simply, these say that for a person to learn, he must be ready or have a desire to learn, his attempt to learn must bring about a positive result for him, and he must repeat an action to help him learn it.

The law of readiness. A person probably won't respond to a stimulus in the way that you would want him to until he is ready to respond. So if you tell a new hand to do something but he isn't giving you his full attention or isn't particularly interested in what you are saying, he probably will not respond to your directions in the way that you would like.

To make sure that a person is ready to respond, or ready to learn, you must first get him to *want* to learn. You do this by getting his attention, arousing his curiosity, and getting him interested in what you want him to do. To simplify this first law of learning, we could say that a person will not respond to training until he wants to respond, or until he is in the right frame of mind to respond.

The law of effect. If the result of responding to a stimulus is pleasing or satisfying, then a person is likely to respond again under similar circumstances. So if a new hand feels successful and proud after learning to perform a job, he will probably be enthusiastic about learning other jobs as well. The law of effect, then, states that if the effect of responding to a stimulus is pleasing, then the response will be repeated.

The reverse of this law is also true. That is, if a person responds to a stimulus and the result is annoying or unpleasant, then he will avoid responding again. If a new hand tries to learn a job and fails, he will tend to avoid learning other jobs.

The law of exercise. When a sensory nerve is stimulated and the brain responds, a connection is made in the body's central nervous system. Each time that stimulus-response action happens, the connection is strengthened. This principle is the basis for the third law of learning, namely, that the more times an instruction is repeated, or exercised, the more likely it is that it will be learned. School childen, for example, learn their multiplication tables or memorize poems by repeating them many times. What they really are doing is exercising the connection until it is so strong that the connection is automatic.

According to the law of exercise, repetition aids in learning.

WHAT ARE THE TRAINING GUIDELINES?

Four basics apply to most training situations

Although teaching methods vary greatly, and no single one is best for all kinds of training, four guidelines apply to most situations. Based on the laws of learning, the four guidelines are (1) get the worker ready to learn, (2) tell *and* show how the job should be done, (3) let the worker do it, and (4) follow up.

Get the worker ready to learn

People who want to learn are the easiest to teach, as the law of readiness states. It is important to get the worker interested in what you are about to teach him—to make him really want to learn. One way to do this is to let the worker know why the job you are about to teach him is important, and why it must be done right. Say, for example, that you are about to teach a new floorman how to load a grease gun. Before you start, explain to him that he will be using a grease gun a lot, since just about everything on the rig that moves needs greasing. If the grease gun isn't loaded right, he'll have a hard time doing his job.

Another way to get the new worker ready to learn is to spend some time getting to know him. Find out a little about him personally, and in particular about his past experience. Not only will this help you to find out what the worker already knows about the job, but it will also help build up the worker's self-confidence. If the worker feels that you respect him, he will show you respect as well. Confidence in you will put him in the right frame of mind for learning the job.

Tell and show how the job should be done

Teaching a job doesn't mean just telling the worker how to go about it or saying, "Watch how I do it." Do both: tell *and* show the new worker how to do the job. The advantage of this approach goes back to the nature of learning discussed earlier. We can learn through all five senses, so the more that are involved, the better.

Before you begin to describe and demonstrate the job, be sure that the employee is standing in a position so that he can see the operation from the same point of view as yours. If he is standing opposite you, he will see everything backwards. This could cause confusion when he later tries to repeat the procedure himself.

Stand side-by-side as you demonstrate a job so that the trainee can see things from the right point of view.

Describe and demonstrate the job step by step, slowly. If too much information is presented too quickly, the worker will probably forget part of it. The law of effect comes into play here: if the worker fails to master a job because too much was expected of him, he will be wary of trying again for fear of repeated failure.

Let the worker do the job

After you've explained and shown the worker how to do the job, it's time to let him try it himself. It is important that you be there to observe the first few trys so that you can see if he's doing everything right. If a worker performs a procedure incorrectly many times, it might be difficult to change later. Remember the law of exercise? Once a connection has been strengthened by repetition, it is difficult to break—even if it's wrong. If a worker does perform a step incorrectly, it is a good idea to have him repeat that step correctly several times.

Follow up

A worker doing a new job has to fly alone sooner or later. So after he has shown you that he can do the work reasonably well while you are standing by, turn him loose for a while. Don't abandon him completely, though. Make a point of checking on his progress and workmanship regularly—perhaps three or four times the first day that he is on his own, then once or twice a day for a week or two. Never assume that he is completely trained. There is always something that he can learn to do, or learn to do better.

An important part of follow-up is letting the worker know that he can come to you if he has any questions about his job. It is a good idea to designate a reliable hand that he can go to when you aren't available.

HOW DO YOU PLAN A TRAINING PROGRAM?

List the jobs, break them down into steps

In the oilfield, as in most work situations, each employee must be trained to do many different jobs, not just one. As used here, the term *job* means any task or assignment that has a beginning and an end and that can be completed in a short period of time. To make sure that the new employee is properly trained in all the jobs that he will be expected to perform, you should track the jobs on an employee progress chart. Breaking down each job into steps will help you during training.

The job list

To begin training a new worker, you should first make a list of all the jobs that the worker will be expected to perform. List the jobs in order of difficulty, beginning with the simplest job and ending with the most complicated and demanding. If the new worker was hired as a floorman, the list of jobs might begin like this:

1. Fill grease gun.
2. Check tongs.
3. Check slips.
4. Measure pipe.
5. Attach hoist line to pipe.
6. Check oil in rotary.
7. Carry mud.
8. Make a connection.
9. Catch a sample.
10. Grease alemite fittings.

This list would continue until it included all the jobs that a floorman performs, ending with the most difficult. Ideally, training would begin with the first job on the list—the easiest—and proceed in numerical order to the end. This is rarely possible, however, because you must choose from the jobs that need to be done at the moment. To be practical, work cannot be interrupted while the new employee is being taught a job.

An employee progress chart (like our sample on page 40) will help you to keep track of a worker's training. The chart should have spaces to list each job that an employee is to learn, then spaces to check off, or sign and date, when you feel that training is complete and the worker has mastered the job. A blank of our progress chart for you to photocopy is provided at the end of the lesson.

A progress chart also can be used to evaluate the competency of employees who have been on the job for awhile, allowing you to identify any areas that need improvement. You might want to review the list of jobs on the progress chart with an older employee so that the employee can point out jobs that he would like to know more about.

EMPLOYEE PROGRESS CHART

Job		Employee					
		JIM	CHUCK	STEVE			
1	fill grease gun	✓	✓	✓			
2	check tongs	✓	✓	✓			
3	check slips	✓	✓	✓			
4	measure pipe	✓		✓			
5	tie catline on pipe	✓					
6	check oil in rotary	✓					
7	carry mud	✓	✓				
8	make a connection	✓					
9	catch a sample	✓	✓				
10	grease alemite fittings	✓		✓			

The job breakdown

After you have listed all the jobs that the new worker will perform, you must then break each job down into separate tasks, or steps. A *step* is defined as an action that causes a job to progress toward completion. Take filling a grease gun, for example—the first job on our progress chart for a new floorman. This job could be broken down into seven steps:

1. Pull out plunger until it locks.
2. Unscrew top of grease gun and remove old cartridge.
3. Remove lid from new cartridge.
4. Insert cartridge into grease gun.
5. Remove pull tab from end of cartridge.
6. Replace top of grease gun.
7. Release plunger.

To break a job down into steps like this, you must think through each motion of the job as it is performed. You will probably want to actually perform the job, jotting down each step as you do it. It is helpful to use a breakdown sheet like the one shown on page 42. (A blank form for you to photocopy is provided at the end of the lesson.)

After you have broken the job down into steps, you should then write down key points for each step. A key point is any information that is necessary to perform the step, or that will help a new worker to do it right. Safety is always a key point. Look at the key points listed on our sample job breakdown sheet for filling a grease gun. Key points are the kind of information that a worker picks up by experience. It is sometimes difficult for an experienced worker to identify key points because he does many of them by habit. But they might make the difference between success and failure for a new worker.

The new employee should not be expected to memorize job breakdowns or key points. The job breakdown form is intended primarily for use by the supervisor as he is training a new worker.

Steps in loading a grease gun

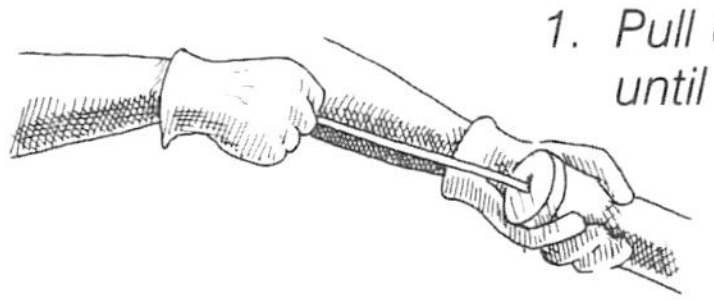

1. Pull out plunger until it locks.

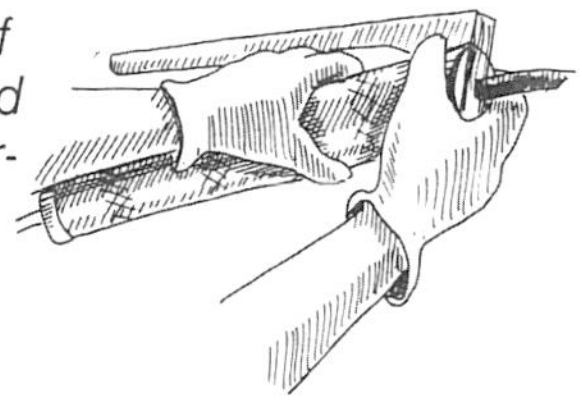

2. Unscrew top of grease gun and remove old cartridge.

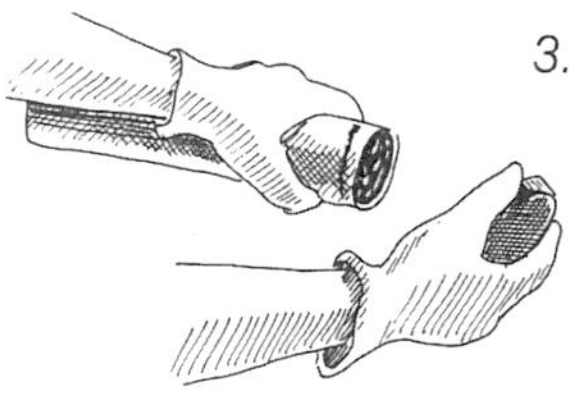

3. Remove lid from cartridge.

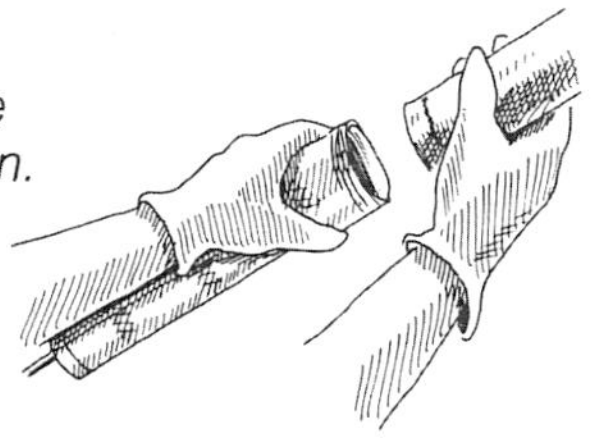

4. Insert cartridge into grease gun.

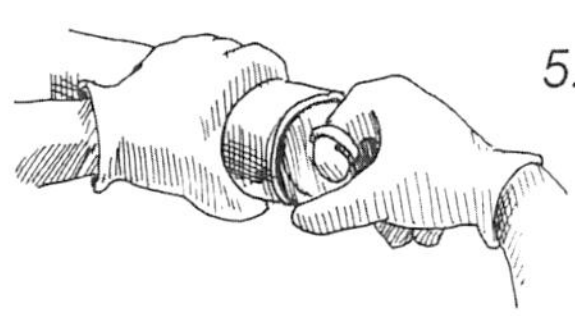

5. Remove pull tab from end of cartridge.

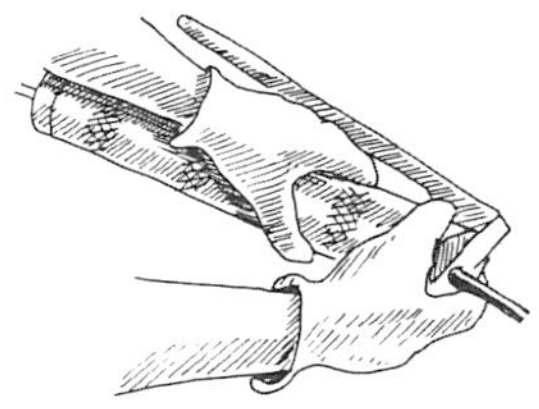

6. Replace top of grease gun.

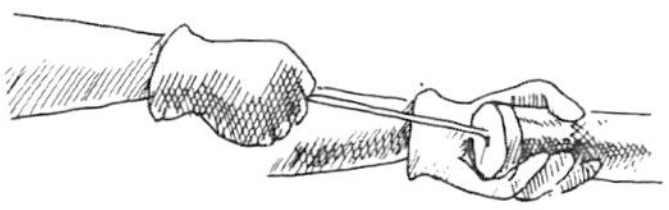

7. Release plunger.

JOB BREAKDOWN SHEET

Job Fill grease gun No. 1

	Steps in the job	Key points
	A step is an action that causes a job to progress toward completion.	Key points are any information that will help the worker do the step right or make it easier.
1	pull out plunger until it locks	make sure plunger is locked open
2	unscrew top of grease gun	be careful not to get alemite fitting in dirt
3	remove lid from cartridge	keep dirt out of grease cartridge
4	insert cartridge into grease gun	make sure you put the right end in first. should be labeled "this end toward plunger"
5	remove pull tab from end of cartridge	if you put the cartridge in right, you'll be able to get to the pull tab
6	replace top of grease gun	don't crossthread it
7	release plunger	
8		

Training diagram

In order of difficulty, the steps in loading a grease gun are ranked as follows (size of circle represents difficulty of each step):

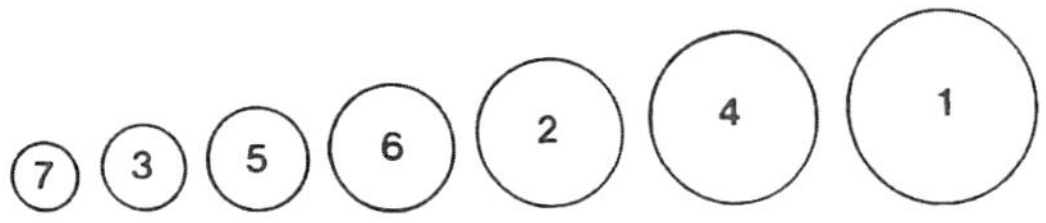

In reality, however, the steps must be taught in the order that they are performed:

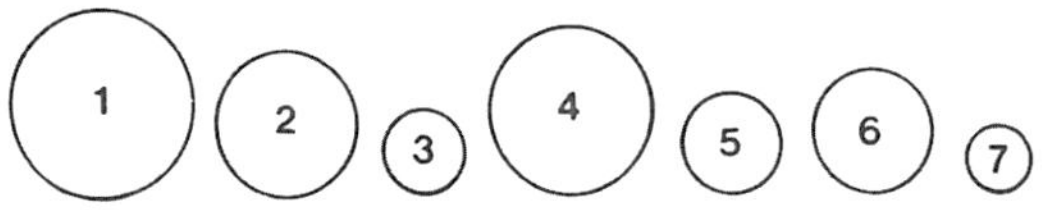

To make training easier, the instructor should allow the trainee to do the easy steps the first few times he demonstrates the job. Each time the instructor repeats the demonstration, the trainee should perform one or two more steps until he has performed all the steps. (Circles represent steps done by the instructor. Squares represent steps done by the trainee.)

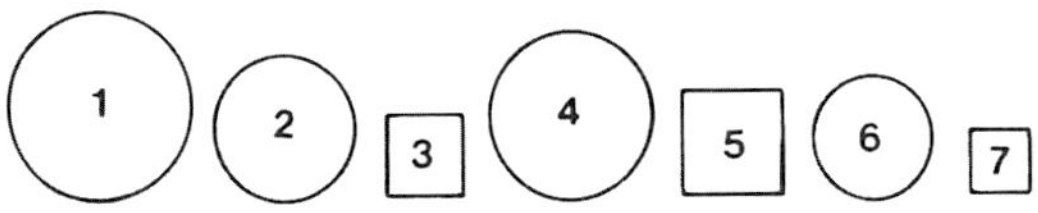

First demonstration

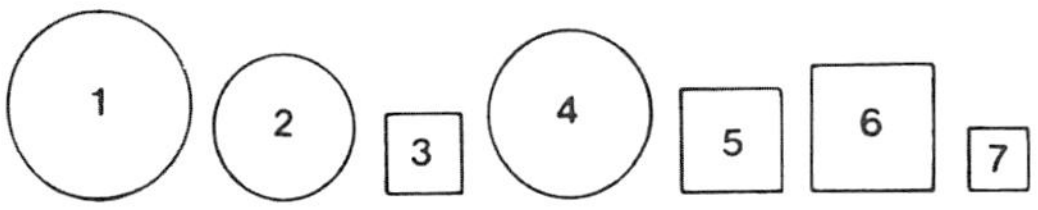

Second

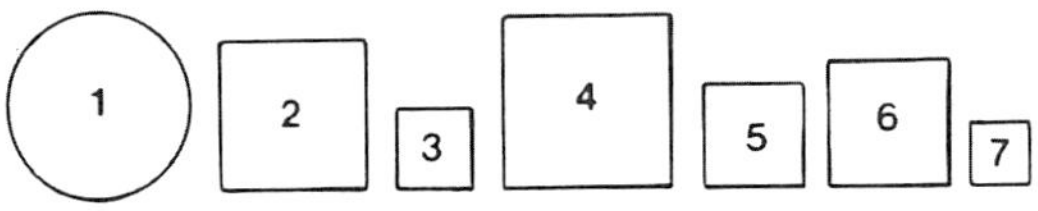

Third

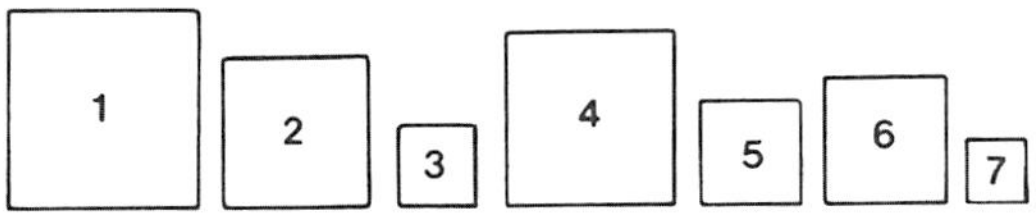

Fourth

Just as it is ideal to teach jobs in their order of difficulty, starting with the simplest and working up to the hardest, it would be convenient if the individual steps in a job could be taught the same way. Unfortunately, steps are almost never arranged in this way for most jobs. Think again about the seven steps in filling a grease gun.

From simplest to hardest, the steps would be ranked as shown in the top diagram at left. There is no way, of course, that the steps could be performed in their order of difficulty. And the worker will have to perform the steps in the normal order when he does the job on his own, anyway. What you can do, however, is let the worker perform only the easiest steps the first few times that you explain the job to him, while you do the harder steps. Each time that you show him the job again, let the worker do one or two more steps, adding steps for him to do in order of difficulty. That way, the employee can work up to the harder steps without doing any steps out of sequence.

Summary

How do you learn? The most basic way that you learn is through the stimulus-response mechanism of the central nervous system. Three laws of learning based on the stimulus-response model are the law of readiness, the law of effect, and the law of exercise. *What are the training guidelines?* Four guidelines that apply to most training situations are (1) get the worker ready to learn, (2) tell *and* show how the job should be done, (3) let the worker do the job, and (4) follow up. *How do you plan a training program?* To begin training a new worker, you should first make a list of all the jobs that he will be expected to perform. A progress chart will help you to keep track of the employee's training in each of the jobs. Breaking down each job into steps will help you during training.

EMPLOYEE PROGRESS CHART

Job		Employee					

JOB BREAKDOWN SHEET

Job ______________________________ No. __________

Steps in the job	Key points
A step is an action that causes a job to progress toward completion.	Key points are any information that will help the worker do the step right or make it easier.
1	
2	
3	
4	
5	
6	
7	
8	

Lesson 6

WHERE DO YOU FIT INTO THE ORGANIZATION?

OBJECTIVES

Upon completing this lesson, the student will be able to

- ☑ *Discuss ways to promote good relations with other first-line supervisors.*
- ☑ *Describe how to improve communications with the boss.*
- ☑ *Explain why staff supervisors are needed and discuss ways to maintain good working relationships with them.*
- ☑ *Describe how to get along well with the customer.*
- ☑ *Discuss ways to improve dealings with service and supply companies.*
- ☑ *Describe ways that a first-line manager can help to promote a good public image for the company and the industry.*

Up to this point, we have mainly been concerned with how a first-line supervisor relates to his crew. But you have to deal with a lot more people than just your crew in order to get the job done. This lesson looks at how you deal with those other people and how you fit into the organization.

As a first-line supervisor, you work with a lot of different people, and you fit right in the middle. In some ways you could be considered a communications center, like a telephone switchboard, receiving messages and sending messages to many different people inside and outside of the company. If you are a driller, for example, you must deal regularly with other drillers, your toolpusher, and staff supervisors inside the company. Outside, you must work with the customer, service companies, supply companies, and the public. How well you handle your relationships with each of these people will play a large part in your success on the job.

HOW DO YOU DEAL WITH OTHER SUPERVISORS INSIDE THE COMPANY?

Tips on how to communicate, cooperate

The keys to making the most of your relationships with other supervisors within the company are communication and cooperation. You're all in this together. So that you and everyone else can do the best possible job, you should make every effort to pass along information that will help other supervisors and try your best to cooperate with them.

Other first-line supervisors

Say that you are a driller. What are the advantages of getting along with drillers on the other tours? Think for a minute about the discussion in Lesson 2 on what people want out of their jobs. One of the most important things is job security—knowing that they will be able to provide for themselves and their families. Since the performance of every employee affects the profits of the company, it is in your best interest to make it as easy as possible for all other employees—the other drillers included—to do their jobs well.

Here are some ways to promote good relations between yourself and other drillers. (The same advice would apply to any first-line supervisors.)

Be considerate. Treat the other drillers like you would want to be treated. For example, don't delay a trip until the next crew comes on tour. Passing off your work to the next guys will only create bad feelings. And always try to leave a clean rig for the next tour. You don't like to clean up after somebody else, and neither do they. Simple things like this can greatly improve teamwork between tours. By making the next driller's job a little easier, you'll also improve the company's ability to make a profit.

Leaving a clean rig for the next tour will promote good relations with other drillers.

Pass information along. Always fill in the next driller coming on about what happened during your shift. Anything to do with safety is especially critical. Say that the safety inspector found a frayed line on the air hoist during your tour. Even though you had it repaired, you should still alert the driller coming on so that he can keep an eye out for any related problems. Such communication can save the next driller many hours of lost time, possible injury, and lost profits for the company.

Offer to help out. If another driller asks you for help, try your best to give it. Even if he doesn't ask, offer help if he seems to need it.

Your boss

The kind of help that you get from your boss depends a lot on the kind of information that you give him. Your boss can't plan or coordinate without full knowledge of what is happening on the job. You must keep your boss informed about your needs and problems. Remember that your problems are his problems.

To approach this a little differently, think of a time when you were in the position of trying to solve a problem brought to you by one of your crew members. You needed detailed information on all aspects of the problem, and you wanted to get it from the person closest to the situation. Only after you had gotten all the facts could you make a decision on how to solve the problem. The same situation exists when your boss is trying to solve a problem, and he will usually look to you for the information that he needs.

Here are some suggestions for how to improve communications between you and your boss:

Keep your boss informed. When you have a problem, go to your boss with all the facts. Do not ask the boss to do your work for you. Make it easy for him to give you a decision. Different bosses will want more or fewer details in reporting. Try to adjust to the wishes of your boss on this.

Make recommendations. After you have thought through a problem carefully, make a recommendation to your boss on how to handle it. Say that you've had a problem with a guy from the personnel office about paperwork. Don't just complain to your boss about him. Think about how to make things better. For example, you might suggest bringing front office people out for a look at the rig so they'll know what you do on your end of the operation.

Support your boss. Once a final decision is reached, accept it as your own and make it work—even if you don't completely agree with it.

Assume full responsibility for your job. Don't put your boss in the position of having to follow up on you and check your work.

Staff supervisors

Why are staff supervisors needed? The main reason is that drilling technology has advanced to the point that no one person can be an expert in every aspect of the job. Think for a minute about the many things involved in a contract drilling operation.

The drilling rig itself represents millions of dollars worth of equipment. Operating and maintaining this equipment properly requires expertise in diesel mechanics, hydraulics, fluids, electricity, electronics, wireline operations, and other areas.

Planning and carrying out the drilling operation involve economics, petroleum engineering, and geology. Financial managers must handle purchasing, record keeping, accounting, payroll, and taxes. Contracts, worker's compensation, and insurance need the expertise of a lawyer.

A first-line supervisor couldn't possibly deal with all of these areas and handle them well. The driller has a full-time job learning all that he needs to know to do his work alone. So staff people, often called support people, are

Financial managers play an important role in a contract drilling operation.

responsible for supplying expertise in their areas when it is needed. A large contractor may have most of these specialists on the payroll. Small companies often use the resources of the operating company, manufacturers, or others.

Staff people are the experts in their fields, so it's to your advantage to develop and maintain good working relationships with them. Here are some ways to do that:

Give accurate information promptly. The accountant, office manager, or whoever can do his job only if the information he receives from the field is prompt and accurate. Remember that you and the staff have mutual responsibilities. Providing the information that they request will help you to do your job better in the long run.

Ask for advice before trouble arises. Consult the specialist while there is still time for him to perform his service. Remember that you are not expected to be an expert at everything.

What can staff people do to get along better with first-line supervisors? Here are a few suggestions:

Explain why. If you request information from a first-line supervisor, tell him why you need it. The more he understands about the reason for your request, the better he will be able to respond.

Visit the work site. Learn all that you can about the work of the first-line supervisor, seeing him in action as much as possible. The more that you know about what he does, the better you will be able to help him when he needs you.

Respond promptly to requests. When a first-line supervisor requests something of you, let him know as soon as possible what you can do and when you can do it. He can then make plans based on your feedback.

Ask for suggestions. First-line supervisors live with problems on the rig every day, and they usually have good ideas on how to deal with them. Ask them for their suggestions.

HOW DO YOU DEAL WITH PEOPLE OUTSIDE THE COMPANY?

Suggestions for keeping on good terms

Although a first-line supervisor mainly works with his crew and with other supervisors inside the company, he must also regularly deal with people from the outside. These include the customer, service companies, supply companies, and the public. By building good relations with these "outsiders," you will help to make sure that future business for the company will run smoothly.

The customer

Since the customer is paying the contractor to drill the well, it is the contractor's responsibility to run the job how the customer wants

You'll get along better with the company man if you try to make him feel like he's a part of the organization.

it run—as long as working conditions are safe. To do otherwise would risk losing the customer. And a satisfied customer is the contractor's best advertisement.

The customer usually is represented at the well site by the company man, who may be an engineer or drilling foreman. Here are a few ways that drillers or toolpushers can get along better with the company man:

Be familiar with the drilling contract. Learn the provisions of the contract and discuss them, when appropriate, with other first-line supervisors and with your crew.

Be friendly. Make the company man feel that he is a part of your organization. This will help you to gain his confidence. Always give him any information that he requests promptly.

Service companies

Two separate interests, the contractor and the operator, are always present at the well site. At times, one or more additional parties are also present, namely, service companies. With so many different interests represented, delays and strained relations are hard to avoid. And the result is bad for everybody—poorer productivity. Rather than try to lay blame, it is better for everybody to make the effort to get along together. Here are two ways that drillers and toolpushers can help make work easier for service companies:

Get involved in planning. Find out in advance about work that is to be done by a service company. Make sure that everybody knows who will do what.

Be prepared. Have the tools, equipment, and personnel ready when the service company gets there. For example, say that a casing crew is going to be doing a job. Talk to the serviceman in advance to find out what kind of power tongs, elevators, or whatever he'll need when his crew gets there. Let him know in advance that he can come to you with any questions or problems.

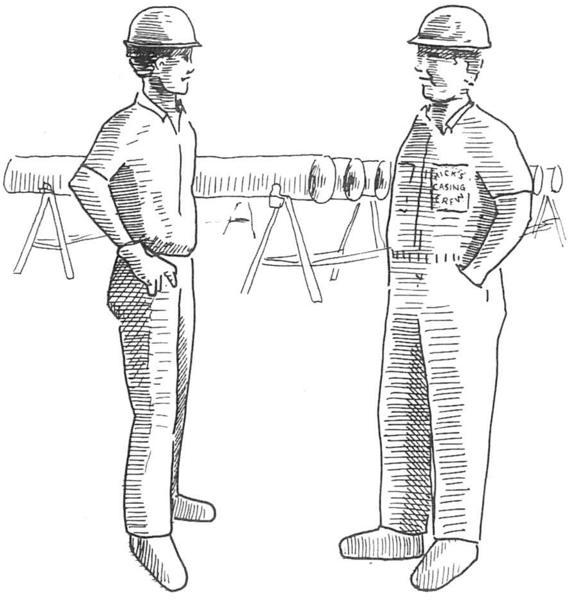

Have tools and equipment ready for the serviceman, and let him know he can come to you with questions.

Supply companies

Supply companies are an important part of the petroleum industry. Some stock as many as 30,000 items, and they must be ready to deliver them to even the most remote drilling sites. Criticism is sometimes directed at supply companies for delivering the wrong part or for not delivering it on time. Often supply companies are accused of overpricing expendable items. As in your dealings with other outside interests, better cooperation and communication will help prevent such problems. A few suggestions toward that end:

Place orders carefully. The most common reason for the delivery of a wrong part is that the order was placed wrong in the first place, or because not enough information was given with the order. Know your equipment, and place orders by name and number.

Anticipate your needs. Place orders early so that you will have what you need well

If a supply company delivers the wrong part, chances are the order was placed wrong to begin with.

before you run out. This will help you to cut back on emergency orders. But don't buy in much greater quantities than you need. This ties up company capital unnecessarily.

Coordinate orders of expendable items. If you order expendable items in large quantities, you can lower transportation costs and therefore unit prices of the items.

The public

Although pipelines, derricks, or refineries are much in evidence in some parts of the country, the only direct contact that most people have with the petroleum industry is when they buy gasoline or other products. Until fairly recently, the average person probably thought little about the industry one way or another. But public relations have become increasingly important to the petroleum industry in recent years.

Price and supply fluctuations and concern over the environmental impact of industry operations have brought the industry under close public scrutiny. And television shows that depict oilmen as robber barons have not helped the industry's image.

Actually, the petroleum business is made up of a few large companies, many smaller companies, and many more hard-working employees. Industry-sponsored television spots and magazine ads attempt to show the industry as a responsible part of the community. But some of the most effective public relation efforts are those by individuals working in the industry. The personal contacts of industry workers with their neighbors and community organizations build more goodwill than almost any other method.

First-line supervisors and rig crews are building public relations, either good or bad, in everyday contacts with people. What can you do to make sure that the impression you make is a good one?

Be a responsible citizen of your community. The opinions of those who know you as an individual are carried over to your company and to the industry as a whole. Likewise, encourage those who work for you to also be good members of the community.

Promote safety on and off the job. Some companies post safety records at the job site in public view. Encourage your crew to take pride in a good record on safety.

A safe work record will help to promote good public relations.

Maintain equipment and property. When the work site looks good, it is a visible sign that things are orderly and well run.

Be courteous. Have regard for others when driving company equipment on public highways. A discourteous person driving a truck bearing the company logo presents a negative image that will prove difficult to overcome.

Summary

How do you deal with other supervisors inside the company? Be considerate of other first-line supervisors, pass information along to them, and offer to help them out when they need it. Keep your boss informed about any problems on the rig, offer possible solutions, and support his decisions. Give staff supervisors prompt, accurate information and ask their advice before trouble arises. *How do you deal with people outside the company?* To get along better with the customer, make yourself familiar with the contract so that you will know what is expected of you, and try to make the customer feel like he is part of the organization. To help make work easier for service companies, get involved in planning and have the necessary tools, equipment, and personnel ready when the service company gets there. When dealing with supply companies, place your orders carefully, anticipate your needs, and coordinate orders of expendable items. To promote good public relations, be a responsible citizen of your community, promote safety, keep up equipment and property, and act courteously when representing the company.

Lesson 7

HOW DO YOU PLAN AND ORGANIZE WORK?

OBJECTIVES

Upon completing this lesson, the student will be able to

- ☑ *Describe the benefits of being an active rather than a passive manager.*
- ☑ *Explain how good planning can help to control costs and prevent accidents on the job.*
- ☑ *Discuss the importance of flexibility in planning.*
- ☑ *Identify the seven steps in planning and explain how they can be applied to scheduling work.*
- ☑ *Write a schedule to plan the work of a crew for a period of time or for a particular job.*

You have probably heard many arguments against planning on the job. The one said most often is that plans never work out like you expect them to, so why bother? Plans do often need to be adjusted, but that's no reason to do away with them. The ability to plan and organize the task at hand in a way that improves worker productivity and increases company profits is what separates good managers from the rest.

This chapter looks at how planning and organizing your work will make you a better manager, in other words, why you should plan on the job. Two areas where planning will really pay off for a manager are cost control and accident prevention. Seven steps in planning have been identified based on a scientific method of problem solving, and these steps can be applied to scheduling the work of your crew. The seven steps are (1) recognize and define the problem, (2) gather all functioning data, (3) evaluate the data, (4) arrange the data in order of importance, (5) consider alternatives and make a decision, (6) take action, and (7) check results. A written schedule will greatly help you in scheduling the work of your crew monthly, daily, or on a particular job.

WHY PLAN?

The benefits of being an active manager

Let's first describe a passive manager, in this case a driller. When he comes on tour, he assembles his crew to make sure that everybody is there. Then he checks in with the driller going off tour to see if all went well during the last tour. Throughout the day, he follows orders from the toolpusher, handles problems as they come up, and orders new supplies as they run out. At the end of his tour, he passes on the current state of affairs to the next driller on duty. Basically, he put in a day—he came to work with no problems and left with no problems.

What's wrong with this kind of managing? It's not so much that it's bad, but it's not good either—just borderline. A driller who puts in a day by just letting things happen probably won't get fired for it, but he won't be promoted either. A main reason that many people are passive managers is to avoid the risk of being blamed if something goes wrong.

A passive manager gives his crew little direction—he just lets things happen.

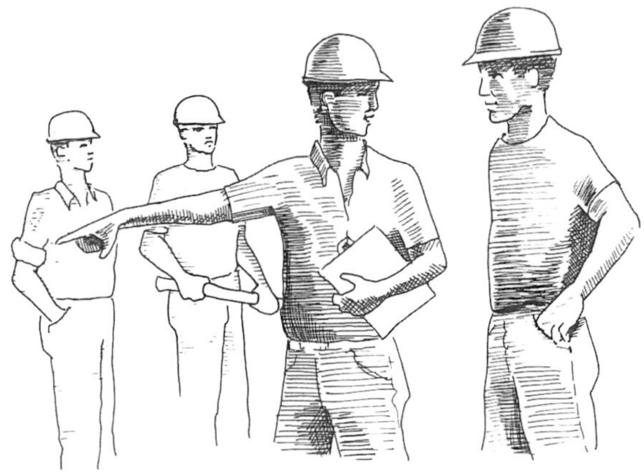

An active manager plans the day for his crew—he sees to it that things turn out like he wants them to.

Compare a passive manager with an active manager. The active manager might be considered an entrepreneur—he's a self-starter who makes things happen instead of just letting them happen. He's always thinking ahead about ways to improve his operation—how to get things done better or faster or more profitably. When the active manager comes on duty, he has a plan for the day. But he's flexible, too, and always ready to revise the plan to make the job go more smoothly. He follows the toolpusher's orders, but he doesn't just wait around to be told what to do. He takes an active part in management, making decisions and then bearing responsibility for them. When things don't turn out as expected, he shoulders the responsibility. And when his plans work out well, he feels a sense of accomplishment from making things happen.

The main difference between passive and active management is planning. The passive manager doesn't plan ahead—he lets the day happen to him. The active manager plans the day and then adjusts as needed—helping to shape the day's events so that they turn out like he wants them to. Two areas where planning really pays off for the active manager are cost control and accident prevention.

Cost control

Planning is a key factor in reducing waste of supplies, abuse of equipment, and loss of man hours—all of which can cut into the profits, and security, of the company. By keeping track of inventory, an active manager can help to cut back on the amount of supplies that are lost or that are ruined by improper storage or exposure to harsh weather. Careful planning can also help to prevent abuse of equipment. And by organizing job assignments so that each worker has something to do throughout the day, an active manager can greatly increase worker productivity.

Accident prevention

The costs of an accident to the company—monetary and otherwise—can be enormous. Direct costs, such as medical expenses, compensation payments, and property damage, are easy to see. But the indirect costs, sometimes not so easy to measure, are usually much higher. Indirect costs include increases in future insurance premiums, wages paid to supervisors or other personnel while their time was required by activities involved with the accident, cost of rescheduling work delayed by the accident, wages paid for working time lost by workers who were not injured, extra cost of overtime work necessitated by the accident, cost of training a new worker, and decreased output of the injured worker after he returns to work. It is even more difficult to place a value on the suffering of the injured worker and his family, loss of customer and public confidence, loss of employee morale, and loss of talent and expertise.

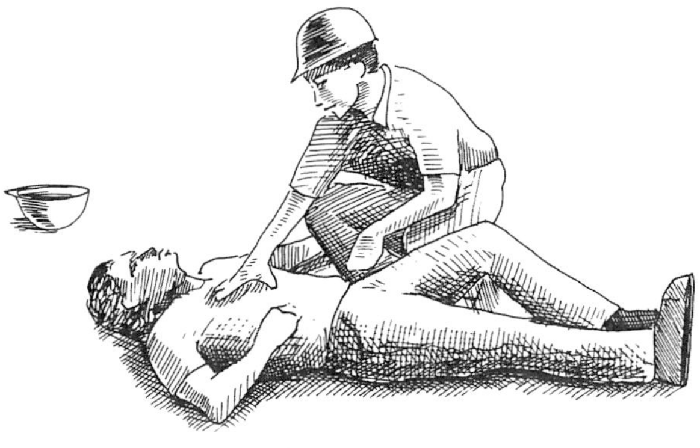

Planning can help to prevent accidents on the rig.

If you think about it, an accident can be defined as an unplanned event. And most accidents on the rig are caused by human error. So good planning by an active manager can greatly contribute to safety on the rig and thereby reduce damage to tools and equipment, reduce possible injury to workers, reduce rig downtime, hold down company costs, reduce loss of time by supervisors and crew, and allow drilling operations to continue.

WHAT IS PLANNING?

The seven steps in planning

Planning is really just an orderly way of solving a problem. The analytical process, also called the scientific method, is an accepted approach to problem solving that can also be applied to planning. It involves seven steps:

1. Recognize and define the problem.
2. Gather all functioning data.
3. Evaluate the data.
4. Arrange the data in order of importance.
5. Consider alternatives and make a decision.
6. Take action.
7. Check results.

Let's take a look at how these seven steps can be applied to scheduling the work of your crew.

Step 1. Recognize and define the problem. Make a list of all the jobs to be done by your crew. You might want to write down each job on an index card, which will make it easy to keep the list up to date.

Step 2. Gather all functioning data. Determine the requirements of each job: number of men, number of hours or days, equipment, materials and supplies, other departments to be involved, and other details.

Step 3. Evaluate the data. Determine the importance and difficulty of each job.

Step 4. Arrange the data in order of importance. Arrange the jobs in order of priority, so that those with the highest priority will be done first. For example, four degrees of priority are (1) an emergency—begin now, (2) do at the beginning of the next work day, (3) must complete by a certain time or date, and (4) do anytime.

Step 5. Consider alternatives and make a decision. Schedule the jobs—decide who will do what and when they will do it. Write the schedule down. Many people use a four-week schedule like the one provided at the end of the lesson. This schedule is really a kind of calendar, called a Gantt chart, listing days of the month down the left side and jobs to be done across the top. (Gantt charts are named for H. L. Gantt, who introduced this type of calendar as a scheduling device.)

Step 6. Take action. Put your schedule into operation. Assign the jobs to your crew members, and advise other departments or subordinate supervisors as necessary.

Step 7. Check results. Check if jobs are being done as scheduled. Few schedules are perfect, so it is almost always necessary to make adjustments. If a schedule works out very well, you might want to file it to use again. Through this practice, procedures can be standardized.

To schedule the work of your crew, start by making a list of all the jobs to be done.

HOW DO YOU PLAN A JOB?

A case study shows how to use a job schedule

The same approach to scheduling work can also be applied to planning a job. The first step is to establish a job description. Once you identify duties and responsibilities, you can assign them on a priority basis. After estimating the time that will be needed for each task, you will then be able to draw up a schedule for the job. For this, you'll need a more detailed schedule than the one used to plan a month's work. The sample job schedule on page 62 allows you to break the job down into hours rather than days. (A blank for you to photocopy is provided at the end of the lesson.) Although similar to an hourly memo on a desk calendar, the hours listed across the top of the chart are hours of job duration, not hours of the day.

To understand how to use this type of schedule, let's consider a sample job—moving a land-based drilling rig. Here are some of the particulars:

- The rig is a National 50.
- The move will cross twenty miles of level ground in West Texas.
- Two crews will be on duty at downtime.
- Blowout equipment will be removed before the beginning of rig down.
- Eight trucks will be available.
- Three five-man crews will be involved in the move.

Take a look at how the job is scheduled on our sample form. (Because this job would take longer than one tour, two forms were used.)

Crews no. 1 and 2 will begin work at the start of the first hour of rig down. Crew no. 1 will clear the floor for three hours, while crew no. 2 will clear the yard and clean the pits. Crew no. 1 will then let down the derrick and remove it, which will require two hours, while crew no. 2 will load the pits and move them to the new location. Loading and moving will require one and one-half hours.

Crew no. 2 will be at the new location at the middle of the fifth hour and begin to reassemble the pits. This part of the job will be completed at the end of the sixth hour. During the sixth hour, crew no. 1 will remove the compound. They will load the substructure in the first half of the seventh hour. Crew no. 2 will be ready to begin laying the substructure at the middle of the eighth hour, and this will be completed at the middle of the ninth hour.

Crew no. 1 will remain at the old location and continue to load pumps, the tool house, and miscellaneous equipment. Going to the second page of our job schedule, you'll see that crew no. 1 will load the derrick at the beginning of the tenth hour and then move the derrick to the new location. They will then begin receiving and raising the derrick at the middle of the eleventh hour. The eleventh hour is the end of the first day.

At the beginning of the second day, all three crews will be at the new location. Crew no. 1 will begin spudding in at the beginning of the eighteenth hour of the job. At the start of the nineteenth hour, crew no. 3 will begin drilling. At the end of that hour, crews no. 1 and no. 2 will go home. The total time for the move is estimated at nineteen hours, according to this schedule.

A simple schedule like the sample on page 62 can help you plan a job.

JOB SCHEDULE

Project ____________________ Starting date ____________________

HOUR / DATE	1	2	3	4	5	6	7	8	9

CREW No. 1

Clear Floor → Let derrick down & remove → Remove compound → Load sub-structure → Load motors, compound, drawworks, pumps → Load tools & Misc. Equip.

CREW No. 2

Clear yard & clean pits → Load pits & Move → Reassemble pits → Lay mats & level → Lay sub-structure

JOB SCHEDULE

Project ____________________ Starting date ____________________

HOUR / DATE	10	11	12	13	14	15	16	17	18

CREW No. 1

Load Derrick → Move derrick to new location → Receive Derrick & Raise → Rig up derrick → Rig up floor → Drill Rat-hole & mousehole → spudin → Go Home

CREW No. 2

Install motors, draw-works, compounds & pumps → Rig Up → Go Home

CREW No. 3

RIG up → Begin Drilling

END OF FIRST DAY ↑

Keep in mind that this schedule only applies to one particular move. If any of the circumstances were different, the schedule might need to be adjusted. And any schedule must be flexible enough to accommodate changes as they come up. If part of the job takes longer or less time than planned, you'll need to shift the schedule to account for that.

You could use the same form that we've used for the rig move to schedule the work of your crew for a day. The eight rows could be used for assigning work for up to eight men. The hours across the top would be the hours in the work day. The objective in planning out the schedule would be to have each man doing productive work as much of the time as possible. You'll have no trouble filling up the spaces during busy times, such as when you're making a trip. During the slow times, you might schedule things like preventive maintenance, housekeeping, or preparing tools for the next use.

Summary

Why plan? Planning is the difference between an active manager, who makes things happen, and a passive manager, who just lets things happen to him. Good planning will help a manager to control costs and prevent accidents on the job. *What is planning?* Planning is an orderly way of solving a problem. The seven steps in the scientific method of problem solving also apply to planning: (1) recognize and define the problem, (2) gather all functioning data, (3) evaluate the data, (4) arrange the data in order of importance, (5) consider alternatives and make a decision, (6) take action, and (7) check results. *How do you plan a job?* A particular job can be planned in the same way that work is scheduled. A written schedule is very helpful but should be flexible enough to allow for unexpected changes.

FOUR-WEEK SCHEDULE

M											
T											
W											
T											
F											
S											
S											
M											
T											
W											
T											
F											
S											
S											
M											
T											
W											
T											
F											
S											
S											
M											
T											
W											
T											
F											
S											
S											

JOB SCHEDULE

Project ______________________ Starting date ______________

HOUR / DATE	1	2	3	4	5	6	7	8	9

Lesson 8

HOW CAN WORK BE DONE MORE EFFICIENTLY?

OBJECTIVES

Upon completing this lesson, the student will be able to

- *Explain why improved efficiency benefits the company and each worker.*
- *Discuss how improved tools can increase job efficiency.*
- *Discuss the role of training in improving job efficiency.*
- *Explain the steps involved in setting up an efficiency program.*
- *Describe how to investigate a problem area where efficiency could be improved.*
- *Define the term* payout period *as it relates to improving job efficiency.*

A profit-making enterprise provides jobs for people. A money-losing enterprise is only on its way to going out of business. Breaking even is not enough. For a company simply to hold its own position in the industry, it must grow. This lesson looks at one of the most important ways that a company improves profits, namely, by operating more efficiently.

Profits and growth mean job security, so improving efficiency on the job is in everybody's best interest. Every supervisor, from first-line on up, can help to improve job efficiency. The two basic ways that efficiency is improved are by making better use of tools and making better use of people. To develop an efficiency program for your operation, you must define the problem area, develop priorities, set goals, develop program objectives, set deadlines, and record progress. Suggested steps for solving a particular efficiency problem are (1) select the problem for study, (2) get the facts, (3) weigh and decide, (4) take action, and (5) check results.

If the solution that you propose will require a change in normal procedures or will cost money, then you must justify the solution by estimating the payout period and the period of improved profit. To present a proposed change to upper management, you should identify the problem, recommend a remedy, estimate the costs and payout period, and outline a plan of action.

WHY SHOULD YOU IMPROVE EFFICIENCY?

Profit and growth mean job security

As a supervisor, you must always be concerned with improving the efficiency and profitability of your operation. The ultimate goal of a business is profit and growth. Profit and growth lead to job security. By looking for ways to improve your organization, you can help to create stable employment and the opportunity for career advancement.

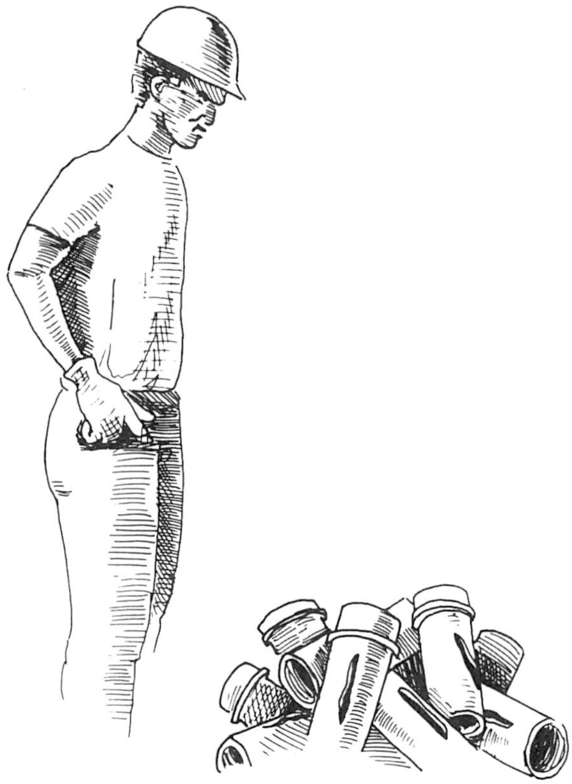

Too many liner washouts will frustrate the crew and cost the company money.

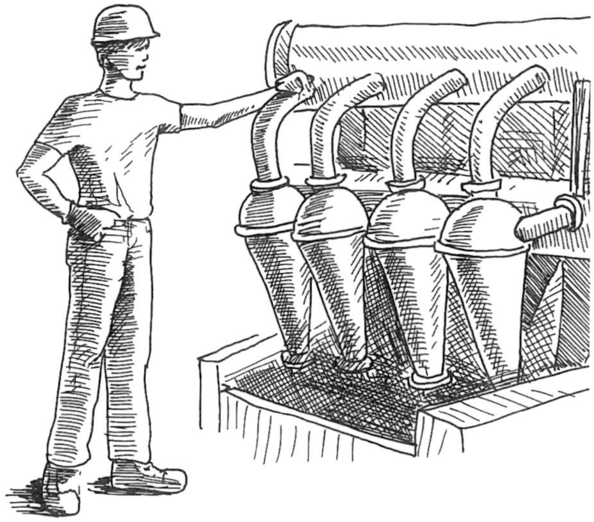

The expense of installing a desander is quickly recovered.

Cost reduction is a term you may hear in connection with improving the financial condition of your operation, but it is a little misleading. Your objective should be not so much reducing costs as improving profits. To improve profits, you must spend money wisely and in a way that produces better results. It can actually require an increase in costs. For example, money spent for a rig capable of drilling wells deeper than the rigs that your company already owns may be a wise investment if a demand for deep wells exists.

An initial outlay can often pay off quickly by reducing labor and material costs, thereby proving more cost efficient than holding back on spending at all. Consider the case of a derrickman who is having trouble with mud pump liners. Suppose that a high sand content in the mud is causing too many liner washouts. As a result, a lot of money is being spent on fluid-end parts and on the labor involved in replacing them. Now suppose that a desander is installed. The expense of installing the desander would be quickly recovered because the labor and material costs would be reduced. Also, the derrickman and other crew members would be free to attend to other duties.

HOW DO YOU IMPROVE EFFICIENCY?

Making better use of tools and people

Man has always tried to find ways to make work easier and to improve his standard of living. One way he has found to do this is by improving tools. In the drilling industry, for example, equipment manufacturers continue to produce improved drilling tools that increase job efficiency. A few examples: spinning tongs or power tongs, which eliminate the need for a spinning chain; air hoists, which reduce the use of catheads for moving some loads on the rig; air-powered slips, which eliminate the need for workers to pull and set manual slips and thereby reduce potential injury to workers and damage to drill pipe; and kelly spinners, which eliminate the need to use a chain to spin up the kelly.

A second way to improve efficiency is by making better use of human resources. That means putting your crew to work in an efficient and effective way. Training plays an important role in helping people to perform at their top capacity. Today, people in the work force are more skilled than ever before. That's because they are better educated than ever before. However, some members of the work force have received a limited formal education. In such cases, on-the-job training and industry-sponsored training courses are offsetting the lack of traditional schooling. In order to increase your knowledge and improve your skills, you should take advantage of all training opportunities. As a supervisor, you should enthusiastically support the participation of your crew members in such programs.

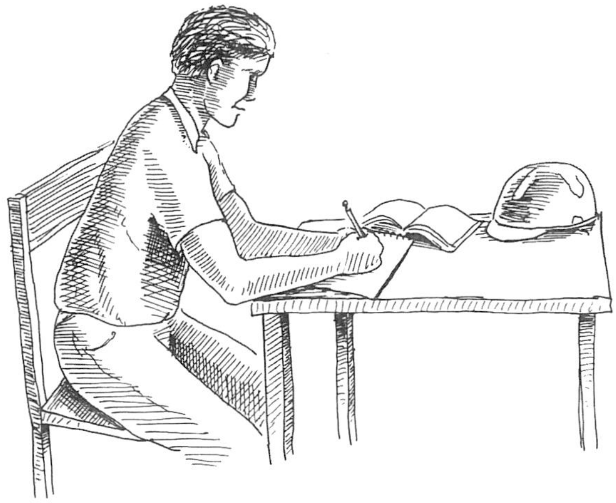

Training helps a company to make more efficient use of its most valuable resource—its people.

HOW DO YOU SET UP AN EFFICIENCY PROGRAM?

Develop a plan and investigate problems

Regardless of the level at which you are a supervisor, you can help to make your operation run more efficiently. To start, you should first set up an efficiency program and then carefully investigate problem areas that could be improved.

The program

The approach to setting up an efficiency program is very similar to the way you plan the work of your crew (discussed in Lesson 7). The following steps are suggested in developing the plan: define the problem area,

Make a list of activities that appear to be wasting money or manpower, or that could lead to injuries.

develop priorities, set goals, develop program objectives, set deadlines, and record progress.

Define the problem area. Take out your pipe book and list any deficiencies that you uncover in your assignment area. Look for equipment, operating procedures, or personnel activities that are currently causing problems or that have caused problems in the past. In particular, look at areas where excessive money is being spent—for example, where considerable manpower is needed, where safety problems exist, or where repeated downtime has occurred. Make a list of all the problem areas.

Develop priorities. Evaluate the list of problems and develop priorities for action to be taken on them. Head up the list with the jobs that offer the most opportunity for increased profit and safety with the least expenditure of time and money.

Set goals. Establish the desired result of the program and the length of time required to get it. Be realistic in setting the number of goals and the time needed. Front-line supervisors might plan three to six months in advance, while higher-level managers might plan several years in advance.

Develop program objectives. Program objectives are steps in accomplishing the goals you have set. If your goal is to complete a project in three months, the first objective might be completed in one month.

Set deadlines. You must set realistic deadlines that you can and will meet. Deadlines are useless unless you strive to meet them.

Record progress. Record, periodically update, and report to your supervisor the results that have been accomplished during the program.

Investigating a problem

Once you've selected a particular area where efficiency needs to be improved, you will need to investigate it thoroughly. The problem may require equipment changes, personnel training or changes, or improved methods. In any case, every detail must be questioned to determine why change is necessary and how change will help to eliminate the problem or simplify the task. To investigate an efficiency problem, you can use a method similar to the one proposed in Lesson 3 on how to resolve personnel problems.

Select the problem for study. Narrow down the area that you will investigate. You can't change everything at once.

Get the facts. Review the record to try to find out what is causing the problem—faulty equipment, poor procedures, people performance, or whatever. Get opinions from the individuals concerned, and make sure you have the whole story.

Weigh and decide. Fit the facts together. Examine each fact to determine its bearing on the problem. Consider possible alternatives and the effect of each on personnel, equipment, and performance. If the change is not within your authority, present the justification to your boss.

Take action. Decide whether or not you are going to handle the problem yourself. If higher authority is required and you receive approval, check the conditions of approval and make proper arrangements. Plan the timing of your action.

Check results. Set up a timetable for following through on your plan. Watch for changes in output, attitudes, and relationships, and whether your action improved performance.

A supervisor may need to present written justification to upper management for some changes.

HOW DO YOU JUSTIFY A CHANGE?

Payout periods and management approval

If the solution to a problem involves changes in normal procedures, investment of capital, or the addition of personnel, then the change must be justified. You can justify changes by estimating the savings in rig time, materials use, or manpower use. All of these savings can be assigned dollar values, and the time required to recover the investment can be reasonably estimated. The time required to recover the investment is referred to as the *payout period.* The accuracy of the payout estimate depends on your judgment and on how well you handled the available facts.

To put it another way, the payout period is the length of time required to recover the money spent on an improvement project. To understand the term better, consider this example. First, let's say that rig cost is $24,000 per day, which is equal to $1,000 per hour. Assume that the savings from an improvement is four hours per week, which totals $4,000 per week. Now suppose the cost of the proposed improvement is $40,000. By dividing the $40,000 cost by the $4,000 per week savings, you get a payout period of ten weeks.

Following the payout period comes the period of improved profit. It is important to have some idea of how long the change will provide improvement, since the change must have a useful life beyond the payout period in order to be profitable.

Most of the problems that you investigate deal with the way in which a job is done. An improved way of doing a particular job does not usually require approval from upper management. However, equipment purchases may require upper-management approval, depending on the policy of your company. In such a case, a driller, for example, must first convince his toolpusher or rig superintendent of the need for a change. Then the toolpusher or superintendent will provide written justification for the change. The presentation to upper management should be short and directly to the point. It should identify the problem, recommend a remedy, estimate costs, estimate the payout period, and outline a plan of action.

Summary

Why should you improve job efficiency? Greater efficiency means higher profits and growth for the company, which lead to job security and opportunities for advancement. *How do you improve job efficiency?* The two basic ways that efficiency can be improved are by making better use of tools and of human resources. Training plays an important role in increasing worker performance. *How do you set up an efficiency program?* The six steps in setting up an efficiency program are (1) define the problem area, (2) develop priorities, (3) set goals, (4) develop program objectives, (5) set deadlines, and (6) record progress. After selecting a particular problem area to investigate, you should get the facts, weigh and decide, take action, and check results. *How do you justify a change?* To justify a change in normal procedures or a change that will cost money, you must estimate the payout period and the period of improved profit.

ANSWERS TO REVIEW QUESTIONS

LESSONS IN ROTARY DRILLING

Unit IV: Man Management and Rig Management

1. E
2. B
3. A
4. C
5. D
6. F
7. D
8. F
9. C
10. D
11. T
12. B
13. A
14. D
15. C
16. B
17. A
18. B
19. A
20. A
21. D
22. C
23. F
24. C
25. A
26. C
27. A
28. B
29. C
30. B
31. C
32. D
33. A
34. E
35. B
36. D
37. T
38. C
39. D
40. C
41. D
42. F
43. T
44. C
45. F
46. F
47. T
48. F
49. F
50. T